AF806439

SpringerBriefs in Economics

SpringerBriefs present concise summaries of cutting-edge research and practical applications across a wide spectrum of fields. Featuring compact volumes of 50 to 125 pages, the series covers a range of content from professional to academic. Typical topics might include:

- A timely report of state-of-the art analytical techniques
- A bridge between new research results, as published in journal articles, and a contextual literature review
- A snapshot of a hot or emerging topic
- An in-depth case study or clinical example
- A presentation of core concepts that students must understand in order to make independent contributions

SpringerBriefs in Economics showcase emerging theory, empirical research, and practical application in microeconomics, macroeconomics, economic policy, public finance, econometrics, regional science, and related fields, from a global author community.

Briefs are characterized by fast, global electronic dissemination, standard publishing contracts, standardized manuscript preparation and formatting guidelines, and expedited production schedules.

More information about this series at http://www.springer.com/series/8876

Juan José Durante · Rafael Rosillo
Editors

Natural Disasters and Climate Change

Innovative Solutions in Financial Risk Management

Springer

Editors
Juan José Durante
Inter-American Development Bank
Washington, DC, USA

Rafael Rosillo
University of Leon
Leon, Spain

ISSN 2191-5504 ISSN 2191-5512 (electronic)
SpringerBriefs in Economics
ISBN 978-3-030-43706-0 ISBN 978-3-030-43708-4 (eBook)
https://doi.org/10.1007/978-3-030-43708-4

This Springer imprint is published by the registered company Springer Nature Switzerland AG
The registered company address is: Gewerbestrasse 11, 6330 Cham, Switzerland

Introduction: Disaster and Climate Change Risk Financing

Natural hazards are defined as natural processes or phenomena that occur in a populated area and can cause the loss of life and livelihood, damage to health and property, social and economic disruption, and/or environmental impacts. Specifically, and for the purpose of this publication, the term refers to all geophysical and climate-related events. Natural hazards occur in all parts of the world making some regions more vulnerable than others. However, they are not considered as disasters unless they cause a serious disruption of the function of a community. In fact, a disaster is the result of the exposure of a population to a hazard, the population's condition of vulnerability, and their insufficient capacity to reduce or cope with any negative consequences (UNISDR 2009).

While there are uncertainties in the trends at a global scale, in the past 20 years there have been 7,570 natural disasters recorded (Guha-Sapir 2019). The causes of these events are complex, however, 91% of these are climate-related disasters. In 2018, the Intergovernmental Panel on Climate Change (IPCC) warned that global warming is likely to reach 1.5 °C between 2030 and 2052. The rise in temperatures increases climate-related risks for both natural and human systems that could affect people's health, livelihoods, food security, water supply, human security, and economic growth. Risks are even greater for disadvantaged and vulnerable populations, indigenous groups, local communities dependent on agricultural and coastal livelihood, and small island developing states.

This is the case for the Latin American and Caribbean (LAC) region, whose countries are highly vulnerable to natural threats. The information available on natural hazards shows that the number of reported disasters in the region increased close to 62% between 1970 and 2018. This upward trend could be driven by better reporting of events and increased exposure and vulnerability. Nevertheless, reported earthquakes have remained stable, while the number of weather-related events has continued to rise (Peduzzi 2005; Banholzer et al. 2014), which suggests that another major force (i.e., climate change) is affecting the frequency of natural hazards. Among the natural disasters that have occurred within the past decade, the following catastrophes can be highlighted: (i) the 2010 Haiti earthquake that resulted in 320,000 deaths and total economic losses of US$7.8 billion; (ii) the 2010 Chile earthquake that caused more than 500 deaths and the overall economic losses was US$30,000 million; (iii) the 2010–2011 floods in Colombia that left 400 deaths and reported economic losses of US$5,000 million; (iv) the 2012 flood in Buenos Aires, Argentina that killed 52 people and resulted on total economic losses of US$1.3 billion; (v) the 2015 Northern Chile floods that caused the death of 178 people and US$ 1.5 billion on economic losses; (vi) the 2016 Ecuador earthquake that caused 673 deaths, affected 389,511 people, and overall economic losses of US$2 billion; and (vii) the 2017 rains associated with the so-called "Niño Costero" in Peru that caused 200 deaths, affected almost 2.2 million people, and resulted on total economic losses of US$3.2 billion (Guha-Sapir 2019).

In their 2019 Global Risk Perception Survey, the World Economic Forum found that environmental-related risks were identified as main concerns in terms of likelihood and negative impact in the next 10 years. The greatest worry was placed on a potential failure of climate change mitigation and adaptation policies and its interconnection with an increase in extreme weather events. To this date, the gross majority of efforts and resources have been placed in the assistance after a natural hazard occurs. However, as the population continues to grow in disaster-prone areas and assets continue to be accumulated, the financial impact is expected to increase. As climate change introduces new uncertainties for governments and communities, international bodies have called for the development of innovative strategies that include the reduction of disaster and climate risk.

From a policy point of view, the concept of disaster risk reduction is to prevent new and reduce existing disaster risk, while managing residual risk. This includes activities that can help understand the risk, strengthen governance, investment for resilience, enhance disaster preparedness, and "build back better". In particular, promoting risk transfer has become a center of the conversation since they increase access to fast and cost-effective liquidity for people affected by disasters. Risk transfer mechanisms aim to shift the responsibility from the risk of loss and damages, mostly financial, caused by natural hazards to a third party. Insurance and reinsurance are the most commonly used tools for risk transfer, however, in recent years other mechanisms such as catastrophe bonds, insurance-linked securities, contingent loans, weather derivatives among others, have become increasingly relevant in the market.

A number of risk transfer mechanisms have been used by LAC countries with the aim of managing disaster risk while providing the necessary financing. For instance, Colombia created an innovative collective insurance policy and Peru designed the El Niño Index insurance project in the coastal region of Peru. Furthermore, catastrophe bonds have been introduced in some countries to provide coverage against exposure to extreme events, as well as the acquisition of continent credit lines for natural disasters. The most prominent case of a financial tool used to fund post-disaster needs is the Caribbean Catastrophe Risk Insurance Facility (CCRIF), the first regional catastrophe pool that aims to contain the fiscal costs, while closing the liquidity gap after a natural disaster (UNFCCC 2012). Despite this, insurance has been the most common tool for governments and individuals to manage disaster risk. For example, both area-yield index-based and weather index insurance have been implemented by different countries in the region, that aim to provide indemnities to farmers and insuring crops against droughts.

The global insurance market reached a total value of US$4.9 trillion in premiums at the end of 2017 and covered US$138 billion of losses from natural catastrophes. The LAC region represented a market share of 3.4% of the total insurance premiums, where insured damages arising from natural disasters accounted for 8% of direct premiums (MAPFRE 2018). Given the estimated impact of natural hazards in the socioeconomic development of the region, there is still a great market potential for the expansion of insurance. Nevertheless, insuring natural hazards such as earthquakes or floods is much harder than to insure other risks. On the one hand, the extreme natural hazards can affect multiple people in different locations at the same time, making it burdensome for the insurance company to file several claims at the same time. On the other hand, when a natural hazard hits an insurance company might have a hard time getting to the area affected which may complicate the claim process. These can drive up the cost of insurance and make it either unavailable or prohibitively expensive, particularly in developing countries.

Alternative risk solutions such as the use of parametric or index-based insurance can provide alternative solutions for insurance and reinsurance companies to finance or transfer risk in a nontraditional way (Markovic and Harry 2018). While traditional insurance relies on the assessment of the actual damage to payout benefits, parametric insurance or index-based insurance does it based on a predetermined, measurable parameter (e.g., rainfall level or wind speed) that can be correlated with the loss and the payout amount is fixed in advance. Parametric insurance products have been around since the late-1990s. The first ones were designed by commodities traders and energy companies, who had high-quality weather data and the right tools to model the relationship between cost/revenues and temperatures that could be quantified and "packaged" to trade weather as a commodity.

Currently, parametric insurance has three basic elements: (i) one or more variables that are closely correlated with the revenue or cost; (ii) threshold levels for disbursement; and (iii) the maximum amount of payout that will be made. Their design requires the analysis of available and measurable data that can allow for a proper design of a trigger or index for the policy. The availability of technology like weather station and satellites has made data more attainable and accurate. As hazard modeling continues to improve, the use of parametric coverage is becoming increasingly efficient and affordable in the market. The use parametric or index-based insurance can reduce asymmetries of information since the payouts are based on an objective index. It also reduces administrative costs, compared to those of traditional insurance, since no individual assessments are needed which, at the same time, increases the timeliness of payouts.

Despite the benefits of using alternative risk solutions, there are a number of challenges that need to be considered in their design. For example, in regions like LAC there can be a lack of data quality and availability that can affect the modeling of the triggers. At the same time, this can cause risk correlations issues that can affect the payouts (also known as basis risk) and therefore, affect the accuracy and efficiency of the solution. There are particular limitations in the implementation of parametric triggers in the insurance market of developing countries where there might not be an enabling environment for its implementation. In some cases, the necessary laws and regulations might not have been in place to facilitate their market development. Furthermore, these products can be technically complex and, to raise awareness about the product and promote uptake, a lot of financial and human capital might be required which can increase the costs.

The following documents focus on the use of parametric triggers in the LAC region through the analysis of successful cases. The first chapter will examine the use of parametric contingent lines at a macro level. Using the Inter-American Developing Bank's Contingent Credit Facility, the document will explain the methodology for designing a state-of-the-art parametric index based on the case of earthquake coverage. The second chapter studies how weather shocks affect farmers in Bolivia. Differentiating by climate risk areas and geographic regions in Bolivia, this chapter documents how yields are reduced with extreme climate variations and what farmers do to cope with these risks. The third chapter will touch on the use of indexed insurance, where a study has been carried out to establish a system of protection against loss of grazing land due to adverse weather, under an experimental type of indexed insurance for the sheep farming regions of Maule and Biobio in Chile.

References

Banholzer S et al (2014) The impact of climate change on natural disasters. Reducing disaster: early warning systems for climate change, Springer, 978-94-017-8597-6, pp 21–49

Food Security Information Network (2018) Global report on food crises 2018. https://www.wfp.org/publications/global-report-food-crises-2018

MAPFRE Economic Research (2018) The Latin American insurance market in 2017, Madrid, Fundación MAPFRE

Guha-Sapir D (2019) EM-DAT: the emergency events database. Université catholique de louvain (UCL)—CRED, Brussels, Belgium. www.emdat.be

Intergovernmental Panel on Climate Change. (2014). Climate Change 2014: Synthesis Report. Contribution of Working Groups I, II and III to the Fifth Assessment Report of the Intergovernmental Panel on Climate Change [Core Writing Team, R.K. Pachauri and L.A. Meyer (eds.)]. IPCC, Geneva, Switzerland, 151 pp

Markovic T, Harry S (2018) Marsh & McLennan insights. Parametric insurance: a tool to increase climate resilience. Retrieved July 15, 2020, from http://www.mmc.com/insights/publications/2018/dec/parametric-insurance-tool-to-increase-climate-resilience.html

Peduzzi P (2005) Is climate change increasing the frequency of hazardous events? Environment & Poverty Times, 3, 7 pp. https://archive-ouverte.unige.ch/unige:32663

SwissRe Corporate Solutions (2018) What is parametric insurance? Retrieved September 15, 2019, from https://corporatesolutions.swissre.com/insights/knowledge/what_is_parametric_insurance.html

UNFCCC Subsidiary Body for Implementation (2012) A literature review on the topics in the context of thematic area 2 of the work programme on loss and damage: a range of approaches to address loss and damage associated with the adverse effects of climate change. FCCC/SBI/2012/INF.14, at 3-4, <unfccc.int/resource/docs/2012/sbi/eng/inf14.pdf> <unfccc.int/resource/docs/2012/sbi/eng/inf14.pdf>
United Nations International Strategy for Disaster Reduction (2009) UNISDR terminology on disaster risk reduction. International strategy for disaster reduction. Geneva, Switzerland
United Nations Office for Disaster Risk Reduction (2017) Risk transfer and insurance for disaster risk management: evidence and lessons learned. Review paper for a special session on risk transfer and insurance at the 5th Global platform for disaster risk. https://www.unisdr.org/files/globalplatform/591d4f658e046Risk_transfer_and_insurance_for_disaster_risk_management_evidence_and_lessons_learned.pdf

Contents

Financial Risk Innovation: Development of Earthquake Parametric Triggers for Contingent Credit Instruments 1
Guillermo Collich, Rafael Rosillo, Juan Martínez, David J Wald, and Juan José Durante

Weather Shocks' Impacts on Farm-Level Agricultural Outcomes in Bolivia 15
Sergio Daga

Pasture Loss Indexed Insurance in Chile 41
María José Pro, Ibar Silva, Julia Sanz, Juan Carlos Cuevas, and Isaac Maldonado

Financial Risk Innovation: Development of Earthquake Parametric Triggers for Contingent Credit Instruments

Guillermo Collich, Rafael Rosillo, Juan Martínez, David J Wald, and Juan José Durante

1 The IDB Contingent Credit Facility for Natural Disaster Emergencies

Latin America and the Caribbean (LAC) nations' high exposure to natural disasters has historically resulted in major impacts on their economic and social development. Public policies and strategies to better deal with the impact of the disasters are constantly evolving. Disasters can have short term impacts upon public finances (i.e. rise in expenditure and loss of revenue) and longer-term consequences for economic growth, development and poverty reduction (e.g., Benson and Clay 2004).

Recognizing the economic and political costs of budget diversions caused by natural hazards, a lot of LAC countries have instituted a catastrophe reserve fund (Mexico's FONDEN being just one example; Charvériat 2000).

However, financial risk transfer instruments are still infrequently used by governments due to the assumptions based on earlier strategies of the most efficient ways of managing financial disaster risks. In economic theory, financial risk retention by governments can be effective if shared or pooled in an efficient and sustainable manner, and without major hurdles or difficulties. This is generally the case for developed countries, which may be endowed with wide and deep domestic financial markets, and stable access to international financial markets. In contrast, developing countries may not able to adequately finance their losses, nor carry on the necessary relief and reconstruction, solely by their own means.

G. Collich · J. Martínez · J. J. Durante
Inter-American Development Bank (IDB), Washington, D.C, USA

R. Rosillo (✉)
University of León, León, Spain
e-mail: rrosc@unileon.es

D. J. Wald
United States Geological Survey (USGS), Golden, CO, USA

J. J. Durante and R. Rosillo (eds.), *Natural Disasters and Climate Change*, SpringerBriefs in Economics, https://doi.org/10.1007/978-3-030-43708-4_1

Studies conducted by multilateral institutions indicate that, in the last several decades, international assistance payments have been declining and the willingness of donors to finance disasters retroactively has been diminishing (Mechler 2004). Moreover, even with these assistance payments, some losses may remain unfinanced. These financing gaps, which manifest themselves in the inability to cover relief and reconstruction losses in a timely and effective fashion, may create additional and significant developmental losses. For instance, important pieces of public services infrastructure, like roads and bridges, cannot be repaired and put into use again, thus hampering the countries' economic recovery efforts.

In more recent years, multilateral financial institutions have been considering ways to develop innovative financial instruments for natural disaster risk management (Andersen 2002). A strategic approach to catastrophe risk financing and background on the technical and economic underpinnings for proposed disaster-linked financial instruments developed by Inter-American Development Bank (IDB) was presented by Andersen et al. (2010).

The IDB, the main Regional Multilateral Financial Institution for the LAC region, has been in the forefront of natural disaster financial assistance since its creation in 1959. Yet, until 2008, most of this assistance was provided on an ex-post basis. In 2008, realizing the need for increasing the ex-ante financing activities by the region's governments, particularly regarding the initial emergency response phase after a major natural disaster, the IDB began to develop a financial disaster risk management strategy (the "Integrated Disaster Risk Management and Finance Approach") to enable those countries with significant Natural Disaster Risk (NDR) exposure to adequately prepare for with high impact natural disaster events.

The goal of developing a financial disaster risk management strategy is to promote better long-term fiscal planning in the face of natural disasters in the region by helping countries to design and implement a fiscal liability financing structure that matches country resources and optimizes their utilization. This process normally entails a complex assessment of the maximum probable losses attributable to a catastrophic event and the subsequent design and implementation of the related financial structure.

Ideally, the design and deployment of disaster risk financing instruments should take place sequentially, after the maximum probable losses on the set of assets at risk have been determined. However, demands at the time of a natural disaster often dictate a more immediate approach, as there is a need to deal with losses that are ongoing or that will, very likely, occur in the near future. Thus, in practice, the design and implementation of risk finance components are undertaken in parallel with other Disaster Risk Management (DRM) activities.

In light of the diverse conditions and resources of the individual borrowing member countries, the IDB considers it necessary to design country-specific action plans to develop their Integrated Disaster Risk Management Programs (IDRMP). Each of these plans includes a multi-prong financial strategy with three basic types of financial instruments.

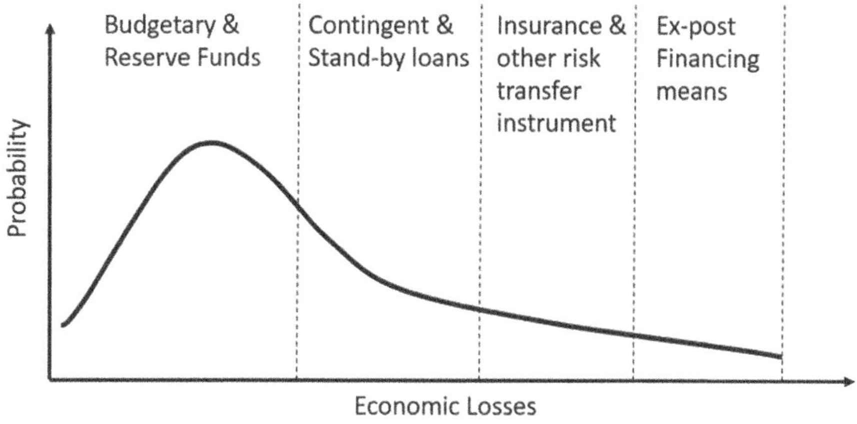

Fig. 1 IDB main risk financial instruments

- Long-term budgetary provisions and reserve funds to cover extraordinary public expenses resulting from highly recurrent natural disaster emergencies (1 in 1 to 5 years recurrence).
- CCF loans for natural disaster emergencies, covering exceedance public expenditures in case of severe, less recurrent events (1 in 5 to 20 years recurrence) and
- Catastrophic natural disaster risk transfer instruments, to cover exceedance public expenditures. In cases of infrequent events of catastrophic magnitude and impact (1 in 20 to 100 years recurrence).

The schematic diagram (Fig. 1) shows the rough coverage sequence of the aforementioned instruments according to the probability of occurrence and the magnitude of the economic losses of the event, and their positioning in the context of a typical hedging strategy for extraordinary public expenses during natural disaster emergencies.

These strategies seek to promote, within the fiscal capacities of each country, a better long-term financial planning for unexpected public expenditure resulting from natural disasters events. This helps countries design and implement a combination of hedging instruments that minimize the financial risk gaps associated with financing and to ensure the efficient allocation of public resources for disaster risk management.

Main Features of CCF Loans

The CCF allows the Bank to structure "Contingent Loans for Natural Disaster Emergencies"[1] (CCF loan) as one of the main financial instruments to support member countries who request it (Andersen et al. 2010). The implementation of management

[1]Stand-by loans that are only available for effective disbursement upon IDB's verification of the occurrence of a natural disaster event of type, magnitude and impact explicitly defined in the related IDB Loan Contract.

strategies follows the Bank's Financial Risks of Natural Disasters. The CCF was approved by the IDB Board of Directors in February 2009 and, as of mid-2019, ten LAC countries benefit from IDB's contingent loans through the CCF. The granting of a CCF loan to an IDB borrowing member country is conditioned on the verification of existence of a comprehensive DRM Program, satisfactory to the Bank, in the requesting country. IDB normally assists the countries with grants for the design of the loan.

The CCF loans extend financial coverage to contractually defined types and intensity of natural disaster events of severe or catastrophic nature. To determine the eligibility for coverage of a particular event, and the amount of funds to be made available for disbursement to the borrowing country under such coverage, the CCF loans utilizes an innovative indexed parametric trigger mechanism based on the type, location, intensity and the estimated number of people directly exposed to the event. For a full description of the range parametric trigger types, see for example, Franco (2015). The calculation of the triggers parameters is based on data provided by reliable, independent sources ("reporting agencies"; Wald and Franco, 2017), acceptable to both the IDB and the borrowing countries.

These triggers allow for a very early, normally within days, verification of the event eligibility for coverage and disbursement of the related loan proceeds. The introduction of these indexed parametric triggers in the operational architecture of the CCF has been a paramount design consideration since, as we discuss later, they are indispensable to ensure both the effectiveness and efficiency of the instrument.

Other important CCF features are the financial conditions of the CCF loans provided through it. The interest rate on disbursed outstanding balances is similar to those applied to regular IDB investment loans. CCF loan's credit fee is paid in arrears, and only on amounts disbursed. But these payments are forgiven if the disbursed resources come from the contractually agreed redirection of Undisbursed Available Loan Balances (UALB) of other Bank loans. There are no other financial charges associated with a CCF loan approval. CCF loan repayment terms are a 25-year final maturity period, repayable in equal semi-annual installments, and an initial principal repayment grace period of five years.

CCF Comparative advantages vis-à-vis ex-post Disaster Financing

CCF loans are capable of providing very timely and conveniently priced financial coverage of extraordinary public expenditures arising in the aftermath of a severe or catastrophic natural disaster event. They thus allow the affected governments to avoid incurring higher financial costs and potential economic and human losses due to lack of sufficient and timely liquid financial resources to take adequate care of natural disaster emergencies. Moreover, timely and efficient coverage of a substantial portion of these unanticipated public expenses that occur during natural disaster emergencies can reduce the negative financial impacts on public finances, potentially stabilize markets, and promote stable economic growth in the medium and long term.

2 Evolution of Parametric Triggers

Inception

Since the beginning of the regulated private insurance industry, two of the most basic tenets have been: (i) policyholders must have an "insurable interest" in order to be able to buy cover to protect their assets (Robertson 2011), should they have no control on the occurrence, magnitude or impact of the insured event; and (ii) the insurance coverage is issued for an amount that do not exceed the value of a total loss of the "insurable interest". These rules have long stifled the expansion of traditional indemnity insurance in certain property and casualty insurance lines, particularly those related to catastrophic events, where the loss assessment and adjustment is particularly difficult to conduct effectively in a timely fashion. This was especially noticeable in the case of natural disaster related coverages, such as those for tropical cyclones and earthquakes.

In the past two decades, the industry began to confront this limitation and to develop new creative ways to overcome them. One of the most innovative methods is the use of parametric coverage products, such as parametric insurance, securities and derivatives[2] that quickly pay out contractually pre-established amounts of funds, upon the occurrence of a clearly defined type of event.

Unlike traditional indemnity insurance products that cover verified insurable losses, parametric risk coverage products pay out in response to defined preset triggers based on the occurrence or intensity of a specified type of covered event. In the immediate aftermath of an insured event, the insurer (or reinsurer) can thus rapidly establish how much it owes the policy/contract holder and provide much faster pay-outs to the insured.

Moreover, the immediate and transparent adjustment made possible by an objective, well-defined trigger system allows parametric policy issuers to provide transparent, uncontroversial, claim settlements in days or weeks, compared to the weeks, months, or even years it may take to finally close traditional insurance claims. Lastly, the use of parametric triggers greatly reduces the administrative cost of adjustment for the insurer, by doing away with the need for individual property inspection and loss claims.

Basis Risk

Parametric coverage is usually cheaper than indemnity (re)insurance because of their lower administrative cost, but the lower prices reflect a greater risk where the policyholder may not recoup their true losses from a disaster. This risk is known in the industry as basis risk (Robertson 2011).

Parametric triggered insurance usually presents higher basis risk than indemnity coverage that, if not adequately mitigated, could end up being costlier for the insured, with the consequent potential negative impact on product demand. The problem was particularly significant with the early trigger models which, primarily

[2]Commonly known as "Insurance-linked securities" (ILS); see Artemis (2016).

due to technological limitations, such as lack of permanent, timely and reliable satellite geo-meteorological observation and information systems, were too broadly designed. These limitations have been substantially reduced through time, as geospatial observation technologies improved, and the industries developed more advanced, accurate, parametric trigger models (with lower basis risks) and the introduction of complementary coverage schemes.

For instance, SwissRe deployed one such new parametric coverage scheme through the Microinsurance Catastrophe Risk Organisation[3] (MiCRO) that offers microentrepreneurs a disaster insurance program in Haiti, developed in the aftermath of the catastrophic 2010 earthquake.

The program aimed to insure (cover) microcredit clients against natural disasters. The innovative feature of this coverage was that, instead of leaving the policyholder bearing basis risk, a sponsored captive[4] insurance company funded by international donors bears it at the insured level, by providing additional loss compensations in cases in which the insurer suffered severe basis risk impact.

Other examples of innovative policyholder basis risk exposures on parametric disaster coverages could be found in the ILS market, where reinsurers enable clients to access lower cost parametric coverages by underwriting reinsurance programs that combine traditional indemnity coverages issuance of diverse ILS, such as CAT Bonds and other structured risk coverage portfolio vehicles, as well as weather derivatives.

Evolution of the Parametric Catastrophe Insurance Triggers

Parametrically triggered insurance covers began to be developed, primarily by larger international reinsurance companies, in the late 1980s and early 1990s both in the catastrophe insurance and catastrophe bond (Cat Bond) markets, where they are still widely used for catastrophic natural disasters. More recently, parametric solutions are increasingly being deployed in public sector and developing markets, as well as being purchased by large corporate clients. The speedy payouts suit both insurance and reinsurance companies, as well as those who may need cash-flow after a disaster. Investors prefer not only the short-term horizon but also the notion they can more easily gauge the odds of paying out parametric insurance (Robertson 2011).

When they first appeared in the international insurance markets, more than two decades ago, the parametric triggers were applied mainly to tropical cyclones and geophysical (i.e., earthquake) events. They were called pure parametric or "cat-in-a-box" triggers, because the cover would only materialize upon verification of occurrence of the covered type and minimum magnitude within a geographically delimited polygon (usually a geographic rectangle, or "box"). In addition, most of these covers were of the "binary" kind. That is, if it was verified that the covered event (type, magnitude and location) occurred, the payout would be 100% of the contracted

[3]MiCRO is a specialty reinsurance company that designs and delivers solutions that enable low-income people to protect themselves against losses after catastrophes.

[4]A captive insurer is an insurance company that is wholly owned and controlled by its insureds; its primary purpose is to insure the risks of its owners. Sponsored captives are captives owned and controlled by parties unrelated to the insured.

cover. If any of the triggering parameters were missed the insurer's payout would be zero. Most covers were on rather reduced areas (normally heavily populated urban areas of high risk, small island states, or subnational jurisdictions), and the coverages were single event (that is, no reinstatement), renewable annually.[5] The information sources for verification of the occurrence of the eligible event were agreed-upon local meteorological or geophysical observation stations or agencies (often referred to as the reporting agencies; Franco 2015).

These limitations seriously limited the type of events that could be covered on a timely, reliable and accurate fashion, usually exposing the coverage to significantly broad basis risk. Moreover, they significantly impacted supply and demand of this type of covers and, therefore, limiting the expansion and deepening of such parametric risk cover markets.

To tackle the situation, the industry progressively introduced improvements in the trigger design and application. Progress was slow at first; it really took off as worldwide permanent and reliable geophysical satellite observations became available. Sources that become broadly available toward the end of the last decade include international space agencies like the National Aeronautics and Space Administration (NASA) and the European Space Agency (ESA), and open, independent earth observation reporting institutions such as United States Geological Survey (USGS) for earthquake notifications, as well as its Prompt Assessment of Global Earthquakes for Response (PAGER), and the National Oceanic Atmospheric Administration's (NOAA) Storm Tracker.

During the current decade, for the purpose of trigger design and operation, the industry was able to practically abandon their reliability on local meteorological stations; offer parametric multi-hazard covers; incorporate new perils, both quick and slow onset perils such as drought, torrential rain and flooding; and move from binary to continuous payout distribution functions.

These type of trigger, dubbed "single parameter trigger", were introduced into the markets with the first Cat Bonds, and were a significant improvement in many ways, yet they failed to sufficiently address the concerns of the industry, the insured, and investors needs regarding basis risk. For this reason, the industry and the markets moved, in part, to multiple parameter triggers, also known as "indexed parametric triggers". A wholesale movement to more complex triggers has not happened since the simplicity of single parametric triggers—despite their high potential basis risk—makes them more straightforward to understand and thus to market (e.g., Franco 2015; Wald and Franco 2017).

The IDB's Indexed Parametric Triggers

IDB developed the Integrated Disaster Risk Management and Finance Approach in 2008 to support the governments of the LAC countries in their efforts to improve their NDR management policies and practices. At that time, IDB adopted indexed parametric triggers for the CCF. The receptiveness by both the governments of the

[5]Multiyear coverages are unusual, mostly because they are more complex to structure and price, and often more expensive due to the policy abandonment risk exposure experienced by the insurer.

region and the international insurance/reinsurance markets was outstanding, to the point that as, of the end of 2019, the bank has developed a portfolio of these parametric products of more than 2B US$, that include operations in ten of its 26-member countries.

The perils covered under this portfolio includes earthquake, hurricanes, floods resulting from torrential precipitation, and volcanic threats. IDB is currently working in the design and implementation of other sudden-onset parametric triggers such as some slow onset events such as droughts and extreme temperature.

All the triggers used by IDB's products to determine the amount of payout are multi-parameter indexed that include at least one parameter related to the intensity of the event. For earthquake contingency loans, the protocol IDB developed follows the shaking intensity distribution generated by the USGS ShakeMap systems (Worden and Wald 2016) and population exposure to such shaking intensities determined by their PAGER system (Wald et al. 2008).

For instance, in the case of earthquakes (from now on EQ) covered under a CCF operation, these parameters are:

(i) the intensity of ground shaking assigned, according to the Modified Mercalli Intensity scale (MMI).
(ii) one vulnerability parameter (based on the percentage of affected population associated with increasing degrees of intensity scale, developed by IDB through actuarial analysis of historical regional EQ casualty and property loss data).
(iii) a parameter on the number of population exposed to different MMI shaking (based on countrywide population distribution estimates, provided by the LandScan global population grid of Oak Ridge National Lab).
(iv) a parameter that reflects the public unexpected expenditures per affected person that the public sector of the covered country expects to have to confront in the event of an EQ of a given intensity.

This last parameter is casuistically determined for each country, together with the government and taking into consideration the expected situation of the country's finances, and its operational capability to timely execute the related procurement during the emergency period, in the immediate aftermath of a severe to catastrophic EQ.

As we discuss in the following sections of this paper, IDB processes all the above described variables of the trigger algorithm, in near–real-time, thus enabling the institution to respond to a country's disaster claim and effectively make the related payout within 72 h of the initiation of the event. To be able to achieve such an impressive turn around, the IDB relays heavily on MMI shaking estimates of the event provided by the USGS, with whom the bank has collaborated in order to improve this specific type of indexed parameter triggering.

3 Trigger Example

Risk financing can complement and stimulate risk reduction. The benefits of financial instruments, including the provision of post-disaster finances for recovery and pre-disaster security necessary for climate adaptation and poverty reduction, are discussed in Linnerooth-Bayer and Hochrainer-Stigler (2015). We present below an example of the methodology designed to execute the CCF Loan in the IDB.

The evaluation of the number of people affected by an EQ is carried out by IDB's team. To perform the necessary calculations, the team uses the USGS PAGER system information, developed to estimate the exposure and impact of an event, and the latest available LandScan Global Population Database, developed by the Department of Energy's Oak Ridge National Laboratory, to estimate the population exposed and affected to each event in question (Wald et al. 2008). A case, particular to post-disaster financing of an earthquake 2016 in Ecuador, with an epicenter 27 km SSE of the city of Muisne, is described.

USGS reported a magnitude 7.8 earthquake on April 16, 2016 at 23:58:36 UTC; its epicenter was at 0.382° N, 79.922° W at a depth of 20.6 km. Figure 2, a portion of the USGS PAGER "onePAGER" summary report, shows contours of the estimated MMI shaking intensity overlain on population density in grey scale. At the time of the event, Ecuador's Ministry of Finance approached IDB to verify the overall eligibility of the event according to what had been established in the Operational Regulations of Ecuador's CCF Loan Contract (Durante 2014). This was the first time the CCF Loan was disbursed by IDB.

The current USGS PAGER summary report aggregates the earthquake impact across the affected region, without consideration of political boundaries between countries. However, since IDB's contractual Natural Disaster Catastrophic Emergency Contingent Loan arrangements differ from country to country, more country-specific information on shaking and exposures was needed to evaluate the effective impact of the damages caused by the event. IDB uses the Geographic Information System (GIS) files provided by USGS in order to calculate the exposed population per country.

According to Ecuador Operational Regulations, the minimum level needed to calculate the exposed population is Modified Mercalli Intensity (MMI) 5.5 and the maximum (MMI) intensity determined for this event was 8.02. Table 1, shows the impact ratio used for computing coverage eligibility (according to MMI intensity) on the exposed population. Using different historical data on the country's impacts from past events, IDB's team developed, in agreement with Ecuador's government authorities, a vulnerability table that establishes a direct relationship between the shaking intensity and the percentage of exposed population that is expected to be significantly affected by the earthquake.

It is important to note that vulnerability tables are included in IDB's CCF Loan Contract with each borrowing country and are described, in detail, in the Operational Regulations section of each contract.

Table 2 shows Ecuador's exposed population as determined by IDB using GIS files

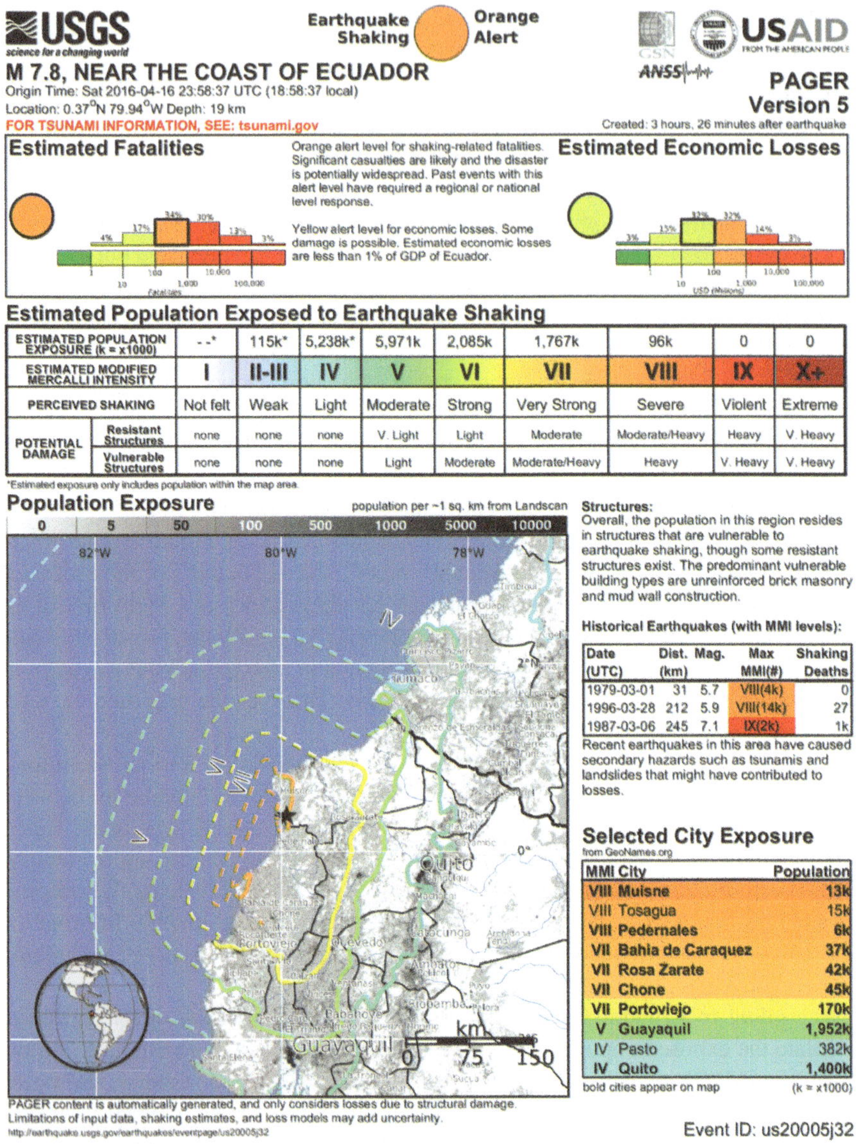

Fig. 2 Top: USGS PAGER summary map for the April 16, 2016 magnitude 7.8 earthquake along the coast of Ecuador. Contours indicated the estimated shaking intensity (labeled with Roman numerals) over a base map of population density (shown in grey scale). The epicenter is shown as a star. Bottom: List of cities and estimated shaking intensity and city population. (Downloaded from https://earthquake.usgs.gov/earthquakes/eventpage/us20005j32/pager; Accessed 08/08/2019)

Table 1 Vulnerability table

MMI scale	Range equivalent to each level of intensity of the scale	Affected population (%)
IV	3.5–4.5	0.0
V	4.5–5.5	0.0
VI	**5.5–6.5**	**12.50**
VII	**6.5–7.5**	**50.0**
VIII	**7.5–8.5**	**85.0**
IX	**8.5–9.5**	**97.50**
X	**9.5–10.5**	**100**

from USGS for the April 16, 2016 earthquake. The information is divided according to the severity of Global Land Survey (GLS) (Gutman et al. 2013). GLS datasets were created as a collaboration between NASA and the USGS from 2009 through 2011. GLS datasets allowed scientists and data users to have access to a consistent, terrain corrected, coordinated collection of data. The left column indicates the MMI scale and the right column indicates the exposed population.

Ecuador's CCF Loan Operational Regulations on Event Eligibility For Drawdowns

It is important to note that some sensitive parameters and/or text have been changed and/or excluded due to confidentiality concerns. The Operational Regulations define the following guidelines and triggers to determine the eligibility for loan disbursements (payouts) upon occurrence of a covered geophysical event.

Earthquakes eligible for coverage are those that occur during the CCF Loan disbursement period. These events have the following characteristics (i) have shown an intensity of VI or greater on the MMI scale, in at least one observation point in the country, according to the ShakeMap reports published by the USGS PAGER system, within a period of 72 h immediately following the onset of the event; and (ii) have affected at least 2% of the country's population within the exposed area.

In order to determine the total population affected by an event, the total populations exposed at each one of the intensity range (Table 2) will be multiplied by the corresponding level in the Vulnerability Table (Table 1). This will allow us to obtain the weight of the affected population at each geographic observation point. The sum of these weighted values will constitute the number of the total population affected by the earthquake. The "Total Population Affected" will then be multiplied by the Covered Population Index (CPI), to determine the maximum amount of IDB's CCF

Table 2 Exposed population at each MMI intensity range estimated by the USGS

MMI (range)	Exposed population (people)
5.5–6.5	2,217,385
6.5–7.5	1,671,345
7.5–8.5	87,709

Loan resources available for disbursement to the country for the specific eligible event.

Ecuador's CCF Loan has been designed to provide liquid resources to finance the country's needs for extraordinary public expenditures during earthquake emergencies affecting at least 2% and up to 6% of the country's total population.

The CPI calculation for ranges of affected population are used as benchmarks for determining the maximum amount of CCF Loan undisbursed resources that IDB would make available to the country if a specific eligible event[6] were to happen. A number for Minimum Population Affected (MiPA) will be established as the trigger value for coverage (which, in this case, has been set at 2% of the population), as well as a Maximum Population Affected (MaPA) as the value at which all the undisbursed and available IDB CCF Loan proceeds would be made available for drawdown by the country (which, in this case, has been set at 6% of the population).

In this sample case, it was determined that the Total Affected Population by the EQ was 7.3% of the country's population. This last number made the 2016 earthquake an eligible event for IDB's CCF Loan drawdown equivalent to the total amount of the loan (300 M US$).

4 Conclusions

This chapter presents one of the financial risk management strategies for natural disasters developed by the IDB, namely the CCF focused on earthquakes. The application of this instrument could be highly beneficial for developing countries. This methodology is used currently to calculate the amount of funds to be made available for disbursement to the borrowing country under such coverage. The CCF loans utilizes an innovative indexed parametric trigger mechanism based on the type, location, intensity and the estimated number of people directly exposed to the event.

The collaboration between USGS and the IDB led to the development of this methodology based on specific earthquake data. The prompt information provided and archived by the USGS after an earthquake in any country, plus the use of indexed parametric triggers, allows the IDB to calculate the total amount of funds to be made available in less than 72 h since the event occurs.

The product is presented with a practical example (earthquake produced in Ecuador on April 16, 2016) of how this methodology is developed. This product is currently used in different LAC countries such as Peru, Ecuador, Dominican Republic, Panama, Honduras and Nicaragua, among others.

[6]Eligible events are defined contractually, based on three parameters: (i) type of event, (ii) intensity of event, and (iii) percentage of the country's total population determined as affected (Affected Population). In this particular example, the minimum requirements are that: (i) the EQ was on the list of covered perils defined in the CCF Loan Contract, (ii) the EQ has reached an intensity of MMI VI in, at least, one observation point in the territory of Ecuador; (iii) and that the number of "Affected Population", as determined by the IDB applying the previously described methodology, is at least equal to 2% of the contractually defined total population of the country.

References

Andersen T (2002) Innovative financial instruments for natural disaster risk management. In: Sustainable development department technical papers series, ENV-140. Inter-American Development Bank, Washington, DC

Andersen T, Marcel M, Collich G, Focke K, Durante JJ (2010) Natural Disasters Financial Risk Management. In: Technical and policy underpinnings for the use of disaster-linked financial instruments in Latin America and the Caribbean, Technical note IDB-TN-175. Inter-American Development Bank, Washington, DC

Artemis (2016) Catastrophe bonds, insurance linked securities, reinsurance capital & investment, risk transfer intelligence. Retrieved May 11, 2016, from www.artemis.com

Benson C, Clay E (2004) Understanding the Economic and financial impacts of natural disasters. In: Disaster risk management series, no 4. World Bank, Washington, DC

Charveriat C (2000) Natural disasters in Latin America and the Caribbean: an overview of risk. Working paper no 434. Research Department, Inter-American Development Bank, Washington, DC

Durante Juan José (2014) Contingent loan for natural disasters emergencies. EC-X1014. Available: https://urldefense.proofpoint.com/v2/url?u=https-3A__ewsdata.rightsindevelopment.org_files_documents_14_IADB-2DEC-2DX1014.df&d=DwIFAw&c=vh6FgFnduejNhPPD0fl_yRaSfZy8CWbWnIf4XJhSqx8&r=VFD7QIE65Y1SMjK7vO8WQIhha7yuh6RgkiwMNhBDCDYhbOQwXaRjxuSIMpzpCiFR&m=JsmbrI-HiekmAPVp1ygQFQOYOi0OO3V4uqIZxQgsKis&s=DTsiALLPkAZBfyjiqwVje-HCU6_fUloVbWZhXBb9ueQ&e=

Franco G (2015) Earthquake mitigation strategies through insurance. In Beer M, Kougioumtzoglou IA, Patelli E, Au IS-K (eds) Encyclopedia of earthquake engineering. Springer, Berlin

Gutman G, Huang C, Chander G, Noojipady P, Masek JG (2013) Assessment of the NASA–USGS global land survey (GLS) datasets. Remote Sens Environ 134:249–265

Linnerooth-Bayer J, Hochrainer-Stigler S (2015) Clim Change 133:85. https://doi.org/10.1007/s10584-013-1035-6

Mechler R (2004) Natural disaster risk management and financing disaster losses in developing countries. Risikoforschung und Versicherungsmanagement, Karlsruhe, Germany. ISBN: 3-89952-120-X

Robertson F (2011) Pay now, argue later. (Winer issue) Insider Quarterly. Insider Publishing. ISSN: 1472-2526

Wald DJ, Earle PS, Porter K, Jaiswal K, Allen TI (2008) Development of the U.S. Geological survey's prompt assessment of global earthquakes for response (PAGER) system. Proc 14th World Conf Earth Eng, p. 8. Beijing

Wald DJ, Franco G (2017) Financial decision-making based on near–real-time earthquake information. Proc 16th World Conf Earth Eng, p. 12. Santiago

Worden CB., Wald DJ (2016) ShakeMap Manual. Retrieved May 10, 2016, from http://usgs.github.io/shakemap/

Weather Shocks' Impacts on Farm-Level Agricultural Outcomes in Bolivia

Sergio Daga

JEL Codes O13; O12; Q12; Q15; Q51; Q54.

1 Introduction

The scientific consensus predicts that without further action against climate change, by the end of the 21st century the Earth's average surface temperature could rise by 2 °C, as the most optimistic forecast, which means reducing agricultural output by almost 25% (IPCC 2014). The rural poor that live in developing countries are more exposed to climate change since their main source of income come from agricultural activities. Another consequence of climate change for farmers is diverting resources from other priorities such as health care or education with negative consequences for human capital and countries' long-term growth and development (Acevedo et al. 2017).

In this paper we study the effects of weather shocks on farmers' agricultural outcomes in Bolivia. A country that in our opinion has not received much attention in the climate-economy literature, in comparison to other countries with same level of development (Dell et al. 2014). Bolivia is one of the poorest countries in South America with approximately 40% of its population living below the national poverty

I would like to thank Alex Armand, Joseph Gomes and Ivan Kim for their collaboration and suggestions. I am extremely grateful with all the staff from Instituto Nacional de Estadística de Bolivia—INE.

S. Daga (✉)
Universidad Privada de Santa Cruz de la Sierra-UPSA, Santa Cruz, Bolivia
e-mail: sergiodaga@upsa.edu.bo

University of Navarra, Navarra Center for International Development, Institute for Culture and Society, Navarra, Spain

J. J. Durante and R. Rosillo (eds.), *Natural Disasters and Climate Change*, SpringerBriefs in Economics, https://doi.org/10.1007/978-3-030-43708-4_2

line. Out of the 3.5 million people that live in rural areas, more than 50% are considered poor; one-third of the labor force work in agricultural related activities, and the agricultural sector is the main source of income for close to three-quarters of the rural population.

Our research is closely related to Aragon et al. (2019) which examines the impact of weather shocks on Peruvian farmers. They study the relationship between extreme temperatures and agricultural outcomes such as yields, total factor productivity, output and input use. They combine micro-level agricultural household surveys with satellite temperature and precipitation records to create a dataset that spans from 2007 to 2015. They apply the data panel approach, which takes advantage of the within-locality variation in weather. They find that extreme heat has detrimental effects on yields and total factor productivity. They also find that farmers increase area planted with extreme heat. They affirm this is a potential evidence of productive adjustment from farmers against weather shocks in order to attenuate undesirable drops in output and consumption.

They also find regional differences in their results. They divide the whole sample between farmers that live in the coast and those that live in the highlands. The coast has a warm semi-arid climate with very little precipitation, while the highland is cooler and receive more rain. On one hand, farmers in the coast tend to be more productive, more educated, and apply more technology such as irrigation. On the other hand, highland farmers are more likely to specialize on low scale farming and less likely to own livestock. Despite coastal farmers in Peru being normally exposed to higher temperatures, the magnitude of the detrimental effect on yields is similar in both regions. However, the increase of area planted in the coast is smaller and less significant than in the highlands. Possible explanations of this result are that availability of land in the coast is more limited, another is that coastal farmers could have other coping mechanisms, or the statistical power of the subsample is low.

Methodologically, this paper follows Deschenes and Greenstone (2007) who argue that with panel estimates it is possible to exploit year-to-year within-district variation in temperature and precipitation to estimate whether agricultural outcomes are affected at the individual level when the year is hotter or wetter than normal. On their study, they find no statistically significant relationship between weather shocks and agricultural profits, corn yields, or soybean yields; they even affirmed that if short-run fluctuations have no impact, then in the long run climate change will have little impact since adaption is possible.

In order to account for nonlinearities in the weather shock function, we take the approach of Schlenker and Roberts (2009). The relationship between agricultural yields and temperature use very fine (1 or 3 °C) temperature bins, polynomials, or piecewise splines. As we do in this paper, they find that yields started to decrease when temperature exceeded the threshold of 29–32 °C interval, depending on the crop. Similarly, they find temperature moderately beneficial for yields when being below the threshold, and harmful above the threshold.

In this paper we find that weather shocks, such as extreme temperature and lack of or excess of rainfall have detrimental effects on Bolivian farmers' yield. We also study the coping actions farmers undertake in order to deal with short-run weather extreme variations. We differentiate our results depending on the risk climate area and the geographic region of the municipalities that the farmers live. On average, farmers that live in highlands municipalities experience a greater reduction in yields. We also observe that farmers living in flood-prone municipalities take specific actions to cope with weather shocks, such as increasing the within farm labor supply in agricultural activities.

Using different data sources, we build up a dataset on agricultural outcomes, farmers' investment and input use. We use a municipality level fixed effects model to evaluate the impacts of weather shocks on agriculture outcomes. We consider potential climate-risk and geographical differences of municipalities. Using the climate vulnerability approach, all municipalities in Bolivia have been categorized accordingly to the most frequent climate risks: flood, drought, and hail or frost. To best of our knowledge, this is the first study that documents the contemporaneous impact of weather shocks on farmers' outcomes in Bolivia.[1]

Our empirical analysis uses the micro-datasets of the National Agricultural Survey for 2008 and 2015. This survey was conducted by the National Office of Statistics of Bolivia (Instituto Nacional de Estadística—INE). Weather data is obtained from high resolution, high frequency satellite data. Weather shocks are constructed following the nonparametric method developed in Schlenker et al. (2006) and Schlenker and Roberts (2009). We construct two measures of cumulative exposure to heat: Degree Days (DD) and Harmful Degree Days (HDD). DD_t is the total daily degrees above a certain lower temperature bound during the growing season in year t. Same-wise, HHD_t is the total daily degrees above a certain upper temperature bound during the growing season in year t. Following Aragon et al. (2019), in this study temperature bounds are endogenously determined from within the data. Hence, DD and HDD are favorable and extreme temperature shocks, respectively. To build precipitation shocks, we gathered daily precipitation measured in millimeters per day from the highest possible resolution. In order to capture the positive effect of rainfall on agriculture, but also the negative effect of extremely low or high levels of precipitation, in all specifications we include a non-linear function of the average daily precipitation in a municipality during the growing season.

We demonstrate that weather shocks, such as extreme temperature and lack of or excess of rainfall have detrimental effects on Bolivian farmers' yield. Specifically, temperatures that exceed department-specific thresholds decrease yield in between 15 to 19%, depending on the climate risk area that the farm is located. The magnitude of the impact of the excess or the scarcity of rainfall on yield is small but negative. Farmers that live in highlands municipalities seem to be more affected by extreme temperature since their yield decreases in about 36%. In terms of coping actions undertaken by farmers to reduce the impact of weather shocks, we observe that extreme temperatures affect the labor supply decision within the farm. The number

[1]Previous works on socioeconomics impacts of weather realizations in Bolivia take municipalities as the unit of observations using cross sectional data and their aim is to analyze implications of climate realizations not of weather shocks. See Andersen and Verner (2014) and Castro (2018).

of members of the farm dedicated to agriculture increases in 5%, but only in flood-prone municipalities, in the rest, the estimate is not significant but with the expected sign. Similarly, farmers in flood-prone municipalities decrease in around 4% the number of hired workers when they suffer from extreme temperatures.

The reminder of the paper is organized as follows. Section 2 presents the context of our study and determine the climate risk areas of Bolivia. Section 3 describes the data sets we use in our study. Section 4 discusses the methodology, the main specification and the weather shock function. Section 5 discusses the results. Finally, Sect. 6 presents the conclusions of our study as well as the policies implications.

2 Bolivia: Context and Climate Risks

2.1 Context

In Bolivia, agriculture is the main occupation for more than 32% of the labor force. In rural areas, it is the main source of income for more than 72% of the working population. Comparing to the other countries that are also located in the Andean Region of South America (Colombia, Ecuador and Peru), Bolivia has the lowest per capita income, its incidence of rural poverty is the highest, its dependence on agriculture is also the highest, and given its usage of land, its agricultural productivity is the lowest. These cross-country differences are depicted in Table 1.

Bolivia is divided in three regions with very distinctive characteristics at the geographical level, but also at the economic and social levels. One is located at the

Table 1 Selected indicators for the andean region countries

	Bolivia	Colombia	Ecuador	Peru
GDP per capita				
Current int. PPP adjusted	6.953,8	13.829,1	11.474,1	12.529,2
Incidence of rural poverty				
% Rural population	57.6	41.4	35.3	46.0
Employment in agriculture				
% Working population	32.1	16.3	25.3	8.1
Agricultural land				
% Land area	34.8	40.5	22.6	19.0
Agriculture, value added				
% of GDP$	13.2	6.8	10.1	7.8
Cereal yield				
Kg per hectare	1.938,0	3.290,5	3.626,7	4.006,6

Source World Bank Development Indicators—Indicators are the last available

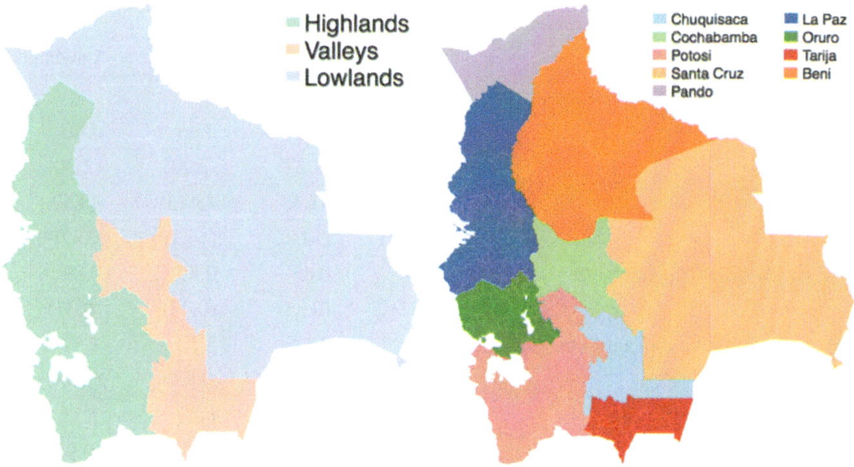

Fig. 1 Regions and departments of Bolivia. *Note* The source of information is the 2012 National Census

western Andean region and covers the country's highlands, with a mean elevation of 3300 masl (meters above sea level). At the other extreme, there are the tropical lowlands, which are located at the northern-east region of the country, with a mean elevation of 600 masl. Finally, the valleys in the south-central portion of the territory form a region lying between high and lowlands, with a mean elevation of around 1800 masl. The highlands region compromises the Departments of La Paz, Oruro and Potosi; the valleys region compromises the Departments of Cochabamba, Chuquisaca and Tarija, and the lowlands the Departments of Santa Cruz, Beni and Pando (see Fig. 1).

Table 2 shows that although average household size in our sample is similar in the three regions of Bolivia, agricultural activities occupy relatively more household members in the highlands and in the valleys than in the lowlands. Coupled with this fact, people in lowlands tend to be more educated and there is a relative higher concentration of poverty—measured by the Unsatisfied Basic Needs Index—among agricultural households both in highlands and valleys than in lowlands.

In terms of agricultural characteristics, Agricultural Productive Units (APU)[2] in the lowlands tend to have higher yields (Kg/Ha), cultivate more land, hire more workers outside the family and invest more in agricultural inputs comparing to APUs in the highlands and valleys. Mechanization and the use of technology is low in all Bolivians APUs, given that the use of tractors and irrigation is only around 30% and less than 20%, respectively, in the whole sample, with no relevant differences among regions.

[2]Agricultural Productive Units (APU) (Unidades de Producción Agropecuaria, UPA) is defined in the 2013 Report of the Agricultural Census as all land that is used totally or partially in agricultural or livestock activities, regardless of size, tenure or legal status.

Table 2 Summary statistics by regions of Bolivia

	All	Highlands	Valleys	Lowlands
Household (HH) characteristics				
HH size	4.31	4.13	4.36	4.40
	(2.27)	(2.27)	(2.23)	(2.30)
HoH age	49.36	51.73	48.43	48.52
	(14.37)	(14.82)	(14.19)	(13.99)
HoH sex (male = 1 female = 0)	0.88	0.85	0.87	0.89
	(0.33)	(0.35)	(0.33)	(0.31)
HH members working in agriculture	2.94	3.12	3.13	2.57
	(1.67)	(1.74)	(1.69)	(1.53)
Years of schooling (19 yo+)	6.86	6.92	6.27	7.53
	(1.52)	(1.51)	(1.68)	(0.92)
Share of poor population (basic needs)	0.67	0.75	0.66	0.63
	(0.17)	(0.17)	(0.19)	(0.14)
Agricultural characteristics				
APU cultivated land (Ha.)	10.37	1.71	2.63	35.31
	(118.89)	(3.70)	(12.16)	(236.97)
APU hired additional workers (yes = 1 no = 0)	0.36	0.26	0.40	0.43
	(0.48)	(0.44)	(0.49)	(0.50)
Number of additional hired workers	3.12	1.84	3.32	4.78
	(9.31)	(4.42)	(9.04)	(13.91)
APU uses irrigation (yes = 1 no = 0)	0.19	0.24	0.22	0.08
	(0.39)	(0.42)	(0.42)	(0.27)
APU uses tractors (yes = 1 no = 0)	0.35	0.37	0.30	0.38
	(0.48)	(0.48)	(0.46)	(0.49)
Total costs of agri inputs (log)	6.27	5.45	6.30	7.28
	(2.15)	(1.83)	(1.99)	(2.35)
UPA average yield (Kg/Ha)	4847.50	3318.85	4219.39	8095.39
	(36075.46)	(16648.76)	(14120.39)	(67321.81)

Note Standard deviations are in parenthesis. *Sources* Agricultural Surveys 2008 and 2015; 2012 National Census; MODIS MOD11C2 module; CHIRPS dataset. The principal cause hindering production is self-reported by the farmer

2.2 *Climate Risks*

According to The Global Climate Risk Index, an index produced by German Watch that analyses to what extent countries have been affected by the impacts of weather-related loss events, Bolivia places as the riskiest country among the ones in the Andean Region (Eckstein et al. 2019). The country is highly prone to extreme weather, often resulting in tremendous loses. In fact, from 1982 to 1983, the negative impact of El Nino resulted in 837 USD million, affecting 2 million people. In 1997–1998, the amount of negative weather shocks accounted for 527 USD million, affecting 166 thousand people. A decade later, from 2006–2008, the loses from both El Nino y La Nina accounted for 956 USD million affecting 1.4 million people. Recently, in 2013–2014, rainfall mainly in the lowlands region causes severe floods affecting 411 thousand people and 384 USD million in looses (Delgadillo and Lazo 2015).

Self-reported assessments about climate risks that regularly hinder agricultural production was collected in the 2015 Agricultural Survey. Farmers declare that the events that affect their production the most are droughts, hailstorm, frost and floods. Taking this as a starting point, to move forward in our analysis we need to identify areas of higher versus lower climate risk, and for that we need to use objective measures of risk.

We build on the work of the World Bank Global Facility for Disaster Reduction and Recovery (Gorriti and Gutierrez 2014) and consider the climatic events reported by farmers to define the risk indicators. The indicator for flood risk considers information such as basin characteristics, drainage, elevation of the terrain, and intensity of precipitation. The indicator for drought risk is instead based mainly on information about aridity and the level of precipitation. The indicators for frost and hail risks are mainly based on terrain elevation, and the correlation between relative humidity and altitude. We then classify the municipalities in Bolivia into Low, Medium and High-risk municipalities based on these criteria. Figure 2 shows the geographical distribution of these risks. We will use this classification throughout this study to understand different responses of farm-level agricultural outcomes to weather shocks depending on the climate risk area it is located.

3 Data

In order to analyze how weather shocks affect farm-level agricultural outcomes in Bolivia, we combine two datasets: agricultural survey and climate data. Our units of observations are the Agricultural Productive Units—APU.

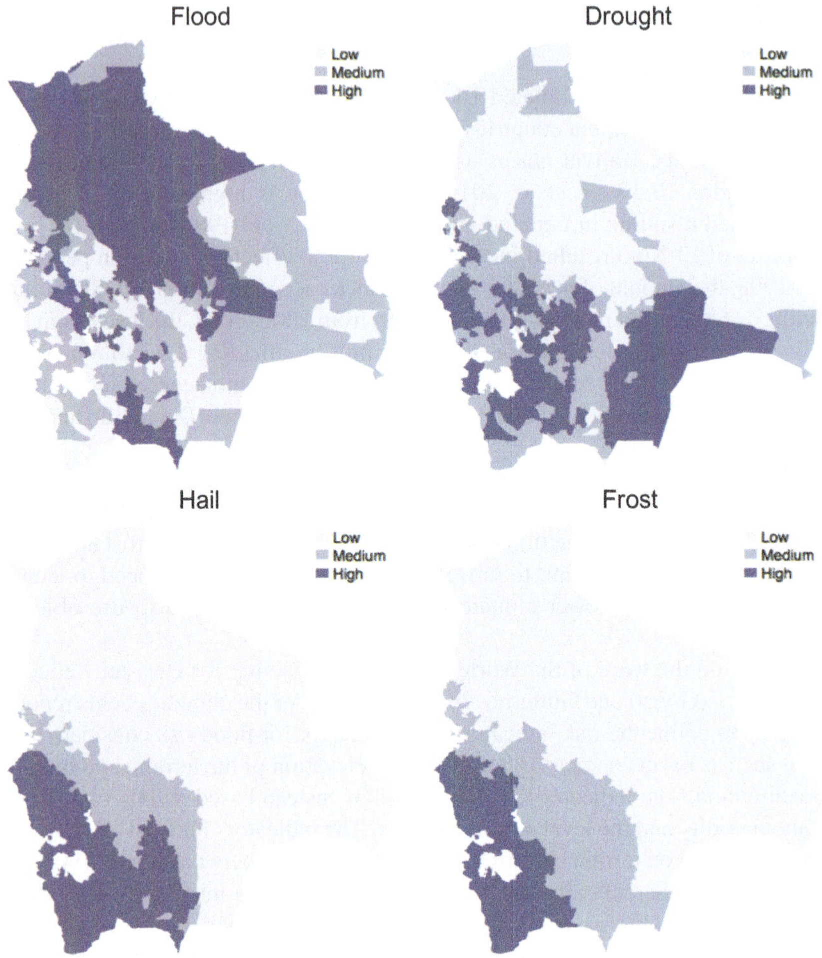

Fig. 2 Distribution of climate risks in Bolivia, by event type. *Note* own elaboration using the World Bank Global Facility for Disaster Reduction and Recovery classification (Gorriti and Gutierrez 2014). Darker colors indicate higher risk

3.1 Agricultural Survey

Farm level outcomes and characteristics are obtained through the 2008 and 2015 Agricultural Surveys, collected by INE. These surveys provide detailed information about farmers' characteristics, inputs and outputs related to agricultural production. In each survey, information refers to the agricultural season starting in July, of the year before, and ending in June, of the year of the interview. In Fig. 3, we show that October and November are the months which the planting occurs the most, while April and May are the months in which harvesting is more intense.

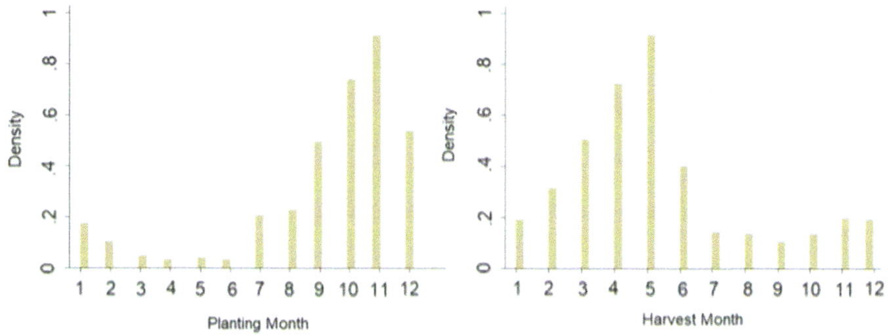

Fig. 3 Planting and harvesting months. *Note* The source of information is the 2008 and 2015 Agricultural Surveys

In Table 3, we provide descriptive statistics from the agricultural survey in which the units of observation are the Agricultural Productive Units (APU) or farms. For APUs that are in the highlands, frost and hailstorm are the most relevant climate events; in turn, APUs that are in the lowlands are exposed to the severe consequences of floods and droughts. This table also shows that irrigation is low in drought-prone municipalities of Bolivia.

3.2 Climate Data

With the aim of building growing season specific weather shocks at the geographically disaggregated level for the whole of Bolivia, we obtain satellite-based weather data. Information about temperature and precipitation. We opted for satellite data instead of weather station-based data because of fewer missing observations.

We obtain Land Surface Temperature (LST) using the MODIS/Terra Land Surface Temperature and Emissivity 8-Day L3 Global 0.05 Deg CMG (MOD11C2) module. The MODIS/Terra Land Surface Temperature and Emissivity (LST/E) products provide pixel level temperature and emissivity values in a sequence of swath-based to grid-based global products. The MODIS/Terra LST/E 8-Day L3 Global 0.05 Deg CMG is configured on a 0.05° latitude/longitude climate modelling grid (CMG).

We first obtain the information at the grid cell level for all of Bolivia and we then average cells at the municipality level. This allows observing, for each day of the year, the average temperature in a specific municipality for both daytime and nighttime. Figure 4 shows the geographical distribution of the average daytime and nighttime LST for the period 2000–2015. Darker colors represent warmer temperatures, while lighter colors represent colder temperatures. We present this division using the temperature distribution in the whole country and for the whole period, and by dividing municipalities into quintiles of the temperature distribution. During daytime, tempera-ture presents a less spatially clustered pattern, with warmer temperatures

Table 3 Descriptive statistics of farmers in the agricultural survey

	All	Municipalities with medium-high risk of		
		Flood	Drought	Frost/Hail
HH size	4.18	4.42	4.41	4.08
	(2.24)	(2.32)	(2.33)	(2.19)
HoH sex (male = 1 female = 0)	0.88	0.90	0.90	0.87
	(0.33)	(0.30)	(0.31)	(0.34)
HoH age	49.28	47.12	47.54	50.18
	(14.35)	(13.39)	(13.94)	(14.59)
Number of additional hired workers	3.12	4.25	3.15	2.84
	(9.31)	(10.06)	(5.52)	(9.57)
Highland	0.35	0.12	0.23	0.44
	(0.48)	(0.32)	(0.42)	(0.50)
Valley	0.38	0.32	0.04	0.47
	(0.49)	(0.47)	(0.19)	(0.50)
Lowland	0.26	0.56	0.73	0.10
	(0.44)	(0.50)	(0.44)	(0.30)
APU cultivated land (Ha)	10.37	29.10	32.67	2.27
	(118.89)	(239.08)	(182.01)	(8.19)
APU uses irrigation (yes = 1 no = 0)	0.19	0.01	0.07	0.26
	(0.39)	(0.12)	(0.25)	(0.44)
UPA average yield (Kg/Ha)	4847.50	8573.18	7297.30	3546.55
	(36075.46)	(50786.68)	(79704.55)	(15707.22)

Note Standard deviations are in parenthesis. *Source* Agricultural Surveys 2008 and 2015

in the lowlands and in the western highlands, and with colder temperatures in the valleys. During night-time, temperature tends to be more spatially correlated and to represent the division in elevation between lowlands, highlands and valleys.

We complement the data on temperature with information on local precipitation. To obtain information about daily precipitation at the highest possible resolution we use the Climate Hazards Group Infra-Red Precipitation with Station data (CHIRPS) database. CHIRPS provide 0.05° resolution satellite imagery supplemented with in situ monitoring station data (Funk et al. 2015).

As we proceeded in the case of temperature, after obtaining the information on precipitation at the grid cell level for all of Bolivia, we average cells at the municipality level. By doing this we can observe for each day of the year the precipitation in a specific municipality. Left panel of Fig. 5 shows the geographical distribution of daily precipitation for the period 2000–2015 in millimeters per day. Right panel presents the average elevation in each municipality measured in meters. As it is observed in both maps, higher precipitation is concentrated in the lowlands.

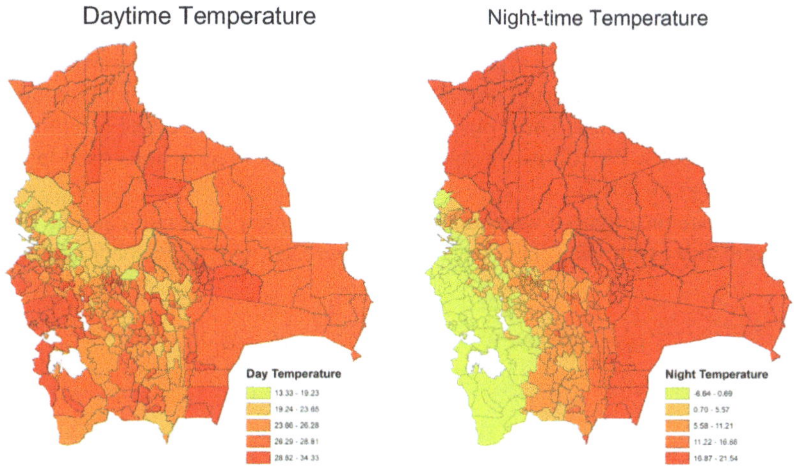

Fig. 4 Land surface temperature (2000–2015). *Note* Each map presents the geographical distribution of municipality-level averages of the corresponding variables in the period 2000-2015. *Source* own elaboration using MODIS MOD11C2 module

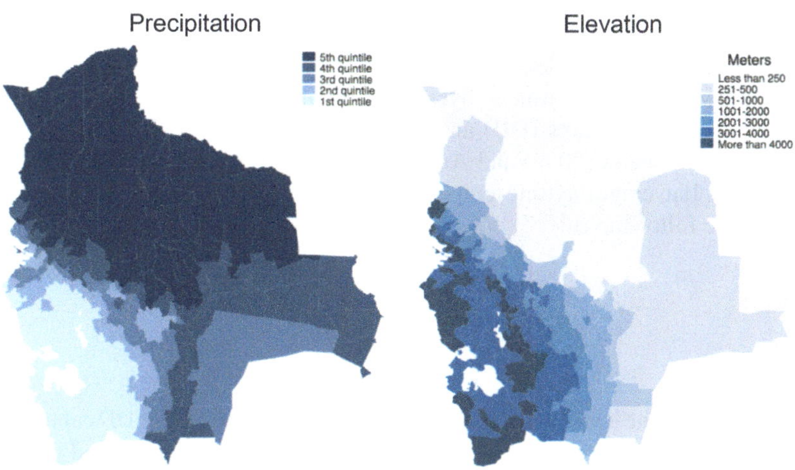

Fig. 5 Precipitation and elevation (2000–2015). *Note* Each map presents the geographical distribution of municipality-level averages of the corresponding variables in the period 2000–2015. *Source* own elaboration using CHIRPS dataset

4 Empirical Strategy

We are interested in measuring how weather shocks affect agricultural outcomes at the farm-level in Bolivia. Thus, our aim is not to estimate the effect of climate but of weather shocks. We follow the approach to the panel regressions used in recent studies (Dell et al. 2014).

We estimate the effect of weather shocks on outcome y of farmer i at time t living in municipality j following the reduced form approach. Our main specification is:

$$y_{itj} = \beta SHOCK_{jt} + X'_{ijt}\delta + \sum_{t=1}^{T} \gamma_t d_t + \alpha_j + \varepsilon_{ijt} \quad (1)$$

where $SHOCK_{jt}$ is the weather shock at time t in municipality j, X'_{ijt} is a matrix of time-varying individual and municipality characteristics, α_j is a municipality-specific fixed effect that absorbs fixed spatial characteristics, whether observed or unobserved, disentangling the shock from many possible sources of omitted variable bias, d_t are time-specific fixed effects that neutralize any common trends, and ε_{ijt} are idiosyncratic error terms, which we assume to be correlated over time for each municipality. Our parameter of interest is β, which captures the effect of changes in weather on individual outcomes.

To capture weather shocks, we build the following indicators of weather shocks that could have influenced farmers' lives. First, we measure shocks related to temperature defining **Degree-days (DD)** and **Harmful Degree Days (HDD)**. Following Schlenker and Roberts (2009) and Aragon et al. (2019) we use a piecewise linear specification. If average daytime temperature for day t is denoted by t_d, we can define DD using the following rule:

$$DD_t = \begin{cases} 0, & t_d \leq t_l \\ t_d - t_l, & t_l < t_d \leq t_u \\ t_u - t_l, & t_u < t_d \end{cases}$$

where t_l and t_u are the lower and upper bound temperature thresholds. DD captures the positive effect of temperature on plant growth and agricultural yields, and indicates the number of degrees in one day that are above t_l and below t_u. If the temperature is below the lower bound, it will account for zero DD, while if the temperature is above the upper bound, it will contribute a fixed amount of t_u minus t_l degree days no matter how high the temperature. On the other hand, HHD measures the degrees in excess from the upper bound capturing the stock of temperature in abnormally warm days. It is computed the same way as DD but using the upper bound as lower bound and positive infinity as upper bound. These specifications of DD and HDD are to distinguish between extreme weather shocks and standard weather shocks. While DD picks up variation in temperature that is not extreme, HDD are instead measuring variation in extreme temperature, which can in turns generate more destructive damage to agriculture.

To capture weather shocks related to precipitations, in all specifications, we always include a non-linear function of the average daily precipitation in a municipality during the growing season. We include a linear and a quadratic term to capture the positive effect of rainfall on agriculture, but also the negative effect of extremely low or extremely high levels of precipitation.

Therefore, the weather shock in Eq. 1 has the following form:

$$SHOCK_{jt} = \omega_1 DD_{jt} + \omega_2 HDD_{jt} + \omega_3 PP_{jt} + \omega_4 PP_{jt}^2. \quad (2)$$

An important issue is to determine the value of t_l and t_u that affect DD_{jt} and HDD_{jt}. For the U.S., Deschenes and Greenstone (2007), Schlenker et al. (2006) and Schlenker and Roberts (2009) set t_l at 8 °C and t_u between 29 and 32 °C, however this is likely to be crop and context dependent and hence might not be transferable to our case.

The geographical and climate risk heterogeneity of Bolivia is also reflected in terms of the variety of cultivated crops[3], therefore, as in Aragon et al. (2019), we do not use estimates of the bounds from the agronomic literature, but we rely on a data-driven approach.

In our sample 90% of the distribution of the daytime temperature in each municipality lies within the range of 5–35 °C. We set the lower bound t_l at 5 °C, hence, the key threshold to estimate will be t_u.[4] We thus create DD and HDD indicators with thresholds starting at 20 °C and increasing by one degree Celsius up to 35 °C. For the whole of Bolivia and for each of its the nine departments, we run 16 separate regressions of yields on our indicators and on average precipitation, at farmer level for the years 2008 and 2015. We then select the upper bounds that maximize the R^2 of these linear regressions. In terms of controls, we also include municipality fixed effects and year fixed effects. Figure 6 suggests that for the whole of Bolivia, 31 °C is the upper bound temperature, which coincides with the one for Cochabamba but not for Santa Cruz, which is 34°C, and is different from the one for Potosi, which is 25 °C. Hence, for each department the upper bound t_u is endogenously determine.

[3]For example, in lowlands, soybeans account for more than 75% of the cultivated crops; in highlands, 30% is quinoa, 25% is potato, 15% is alfalfa, and 10% is corn; while in valleys, 65% of the cultivated land is dedicated to corn, and 20% to potato.

[4]Our results do not change if we include observations below or above this 5 °C bound. In fact, as a robustness check we estimate t_u running different regressions with $0 \leq t_u \leq 10$.

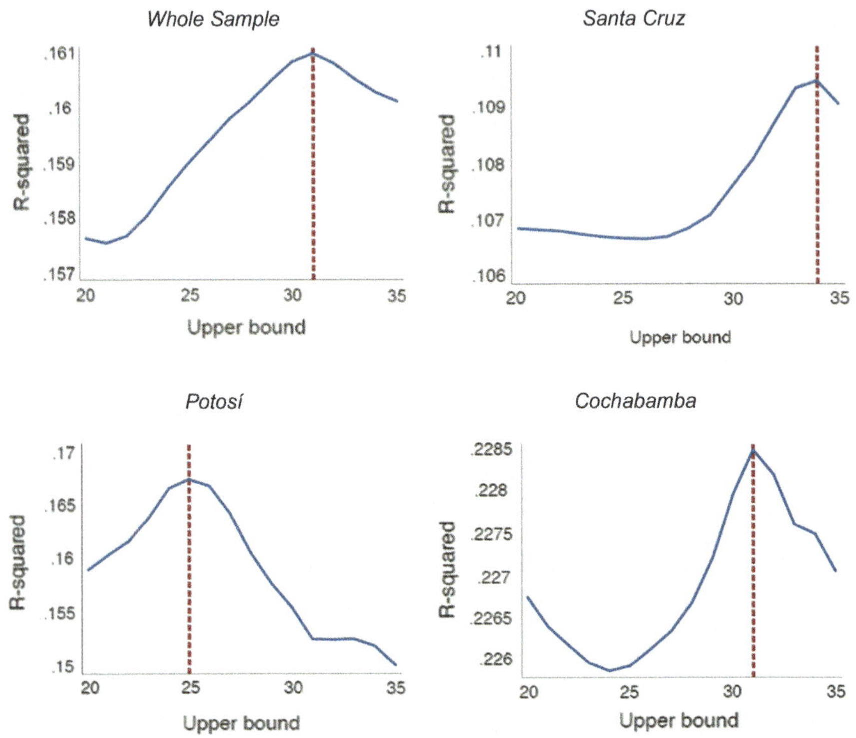

Fig. 6 Model fit (R^2) of weather regressions, by threshold. *Note* Each graph presents the fitness of a model that regresses yields on DD and HDD in order to obtain the upper bound temperature t_u

5 Results

We are interested in understanding how weather shocks affect agricultural productivity at the farm-level in Bolivia. We make use of data on cultivated crops and production for each of the Agricultural Productive Units (APUs) available in the Agricultural Survey for the years 2008 and 2015. The average yield for each farmland is built by averaging crop-specific yield weighted by cultivated area. Yield per farmer is measured in kilograms per hectare and reported in logs with aim of reading the results as percentage increase in yield. The Fig. 7 shows the distribution of yield during the growing seasons of 2007/08 and 2014/15. We can observe that over time, the yield distribution shifted slightly to the right, but yields improvements have been very limited in this period.

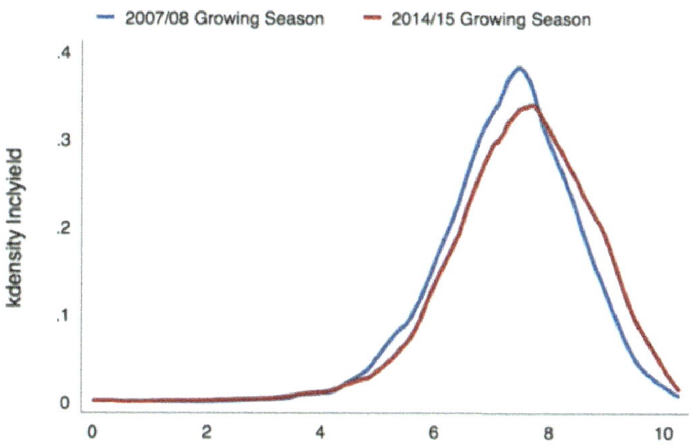

Fig. 7 Distribution of Yields for the 2007/2008 and 2014/2015 Growing Seasons. *Note* Own calculations using the INE's Agricultural Survey of 2008 and 2015

5.1 Effects of Weather Shocks on Yield

The effects of weather shocks on farmers' yield, using our main specification in Eq. 1 and the weather shocks form in Eq. 2, are presented in Table 4 for all farmers in our sample (columns 1 and 2) and also differentiating by the climate risk area in which the municipality of the farmer is located. Similarly, in Table 5 the effects of weather shocks on yield are shown for the whole sample and differentiating by the region in which the farm is located.

For the whole sample, and taking into account additional controls, column 2 of Table 4 shows that extreme temperature (HDD) has statistically significant detrimental effects on farmers' yield of about 15%. This result is repeated for farmers across the different climate risk areas of Bolivia. Similarly, lack of or excess of rainfall has statistically significant detrimental effects on Bolivian farmers' yield of 1.1%. In addition, for the whole sample, normal rainfall increases farmers' yield in around 24%, and up to 34% for farmers that live in drought-prone municipalities. For farmers that live in flood-prone municipalities lack of or excess of rainfall has detrimental effects on yield, but very limited in magnitude. Favorable temperature (DD) increases farmers' yield in around 17% in drought—and frost-prone areas. Finally, Table 5 shows that extreme temperature (HHD) causes a statistically and significant reduction of yield in 36% for farmers that live in highlands municipalities. These results suggest that weather shocks are a particularly important determinants of farmers' productivity and the magnitude of the shocks are relevant.

As robustness checks, Table 12 in Appendix shows the effects of weather shocks on farmers' yield using precipitation above and below 1 Standard Deviation. The results confirm the detrimental effects of extreme weather variations on agricultural

Table 4 Weather shocks and yield, by climate risk areas

Sub-sample	Dependent variable: yield (logged)				
	All (1)	All (2)	Municipalities with medium-high risk of		
			Flood (3)	Drought (4)	Frost (5)
Degree days	0.112	0.110	0.139	0.169**	0.174**
	(0.072)	(0.073)	(0.088)	(0.084)	(0.084)
Harmful degree days	−0.068	−0.152**	−0.184**	−0.175**	−0.191**
	(0.059)	(0.068)	(0.082)	(0.074)	(0.085)
Rainfall		0.239***	0.247***	0.343*	0.320
		(0.076)	(0.088)	(0.193)	(0.247)
Rainfall squared		−0.011***	−0.011***	−0.022	−0.017
		(0.003)	(0.004)	(0.015)	(0.020)
Observations	14635	14562	11011	10941	10908
Municipality FE	Yes	Yes	Yes	Yes	Yes
Year FE	Yes	Yes	Yes	Yes	Yes
Municipality x year FE	Yes	Yes	Yes	Yes	Yes
Additional controls	No	Yes	Yes	Yes	Yes

Note *** p < 0:01, ** p < 0:05, * p < 0:1. Standard errors in parenthesis are clustered at municipality level. The dependent variable is the average yield, reported in logarithm. Additional controls include household head's age, irrigation, and cultivated land. Data stem from two survey rounds done in 2008 and 2015

yields. Similarly, in order to observe whether other outcomes such as the value of livestock owned by farmers could also be affected by weather shocks, Table 13 in Appendix shows that extreme temperature decreases in 19% the value of livestock owned by farmers in Bolivia.

5.2 *Coping with Weather Shocks*

The climate-economy literature suggests that we should not only look at the effects of weather shocks on individuals' outcomes, but also try to understand what actions they are putting in place in order to cope with the consequences of extreme weather variations. Taking advantage of the information available in the Agricultural Survey 2008 and 2015, we examine whether coping with weather shocks for farmers in Bolivia is associated with more land planted, or if there are changes in the supply of labor of the members of the farm or if farms hire more or less workers.

Tables 6 and 7 show that favorable temperature (DD) and normal rainfall have both statistically positive effects of 6% increase and 8% increase of cultivated land by farmers (columns 1 and 2). However, cultivating more land does not seem to be a

Table 5 Weather shocks and yield, by regions

Sub-sample	Dependent variable: yield (logged)				
	All (1)	All (2)	Municipalities in		
			Highlands (3)	Valleys (4)	Lowlands (5)
Degree days	0.112	0.110	0.155	0.108	−0.251*
	(0.072)	(0.073)	(0.107)	(0.105)	(0.148)
Harmful degree days	−0.068	−0.152**	−0.367**	−0.086	0.112
	(0.059)	(0.068)	(0.148)	(0.101)	(0.228)
Rainfall		0.239***	0.979	0.129	0.298
		(0.076)	(0.665)	(0.146)	(0.272)
Rainfall squared		−0.011***	−0.083	−0.007	−0.017
		(0.003)	(0.065)	(0.006)	(0.016)
Observations	14635	14562	5704	5850	3008
Municipality FE	Yes	Yes	Yes	Yes	Yes
year FE	Yes	Yes	Yes	Yes	Yes
Municipality x year FE	Yes	Yes	Yes	Yes	Yes
Additional controls	No	Yes	Yes	Yes	Yes

Note *** p < 0:01, ** p < 0:05, * p < 0:1. Standard errors in parenthesis are clustered at municipality level. The dependent variable is the average yield, reported in logarithm. Additional controls include household head's age, irrigation, and cultivated land. Data stem from two survey rounds done in 2008 and 2015

coping strategy for farmers in Bolivia against weather shocks. This seems to be true for all farmers in the sample, independently where they are located.

Regarding variation in the supply of labor within the farm, on one hand, Table 8 shows that only farms that live in flood-prone municipalities increase the number of members dedicated to agriculture activities in 5% when they are hit by extreme temperature. On the other hand, normal rainfall decreases the number of farm members dedicated to agriculture in farmlands that are in drought—and frost-prone municipalities. The possible explanation is that since normal rainfall is positive for yield, this could permit some members in the APU to look for better paid jobs outside the agricultural sector in order to increase family income. Table 9 shows that the effect just described is statistically significant for farmers in municipalities located in the valley's region. The complement of the findings in Table 8 is shown in Table 10 where we observe whether weather shocks effect the hiring of more workers made by APUs. Indeed, farms in flood-prone municipalities decrease in around 4% the number of hired workers when they suffer from extreme temperature (Table 11).

Table 6 Weather shocks and cultivated land, by climate risk areas

Sub-sample	Dependent variable: Cultivated Land (Logged)				
	All (1)	All (2)	Municipalities with medium-high risk of		
			Flood (3)	Drought (4)	Frost (5)
Degree days	0.060*	0.063*	0.067	0.073*	0.111***
	(0.034)	(0.038)	(0.045)	(0.043)	(0.040)
Harmful degree days	0.034	0.008	0.006	0.014	0.008
	(0.025)	(0.029)	(0.032)	(0.032)	(0.030)
Rainfall		0.079*	0.078*	0.124	0.219**
		(0.043)	(0.047)	(0.088)	(0.107)
Rainfall squared		−0.005***	−0.005**	−0.008	−0.016**
		(0.002)	(0.002)	(0.006)	(0.008)
Observations	15132	15057	11428	11257	11159
Municipality FE	Yes	Yes	Yes	Yes	Yes
Year FE	Yes	Yes	Yes	Yes	Yes
Municipality x year FE	Yes	Yes	Yes	Yes	Yes
Additional controls	No	Yes	Yes	Yes	Yes

Note *** p < 0:01, ** p < 0:05, * p < 0:1. Standard errors in parenthesis are clustered at municipality level. The dependent variable is the cultivated land by farmer, reported in logarithm. Additional controls include household head's age, irrigation, and cultivated land. Data stem from two survey rounds done in 2008 and 2015

Table 7 Weather shocks and cultivated land, by regions

Sub-sample	Dependent variable: cultivated land (logged)				
	All (1)	All (2)	Municipalities in		
			Highlands (3)	Valleys (4)	Lowlands (5)
Degree days	0.060*	0.063*	0.037	0.054	0.063
	(0.034)	(0.038)	(0.054)	(0.055)	(0.063)
Harmful degree days	0.034	0.008	−0.097	0.053	−0.109
	(0.025)	(0.029)	(0.066)	(0.040)	(0.120)
Rainfall		0.079*	0.136	0.062	0.042
		(0.043)	(0.365)	(0.074)	(0.124)
Rainfall squared		−0.005***	−0.030	−0.004	−0.001
		(0.002)	(0.036)	(0.003)	(0.007)
Observations	15132	15057	5818	6067	3172
Municipality FE	Yes	Yes	Yes	Yes	Yes
Year FE	Yes	Yes	Yes	Yes	Yes
Municipality x year FE	Yes	Yes	Yes	Yes	Yes
Additional controls	No	Yes	Yes	Yes	Yes

Note *** p < 0:01, ** p < 0:05, * p < 0:1. Standard errors in parenthesis are clustered at municipality level. The dependent variable is the cultivated land by farmer, reported in logarithm. Additional controls include household head's age, irrigation, and cultivated land. Data stem from two survey rounds done in 2008 and 2015

Table 8 Weather shocks and labor supply within the farm, by climate risk areas

Sub-sample	Dependent variable: number of farm members in agriculture (logged)				
	All (1)	All (2)	Municipalities with medium-high risk of		
			Flood (3)	Drought (4)	Frost (5)
Degree days	−0.002	0.024	0.032	0.030	0.030
	(0.023)	(0.022)	(0.025)	(0.024)	(0.024)
Harmful degree days	0.013	0.024	0.055**	0.013	0.032
	(0.021)	(0.021)	(0.025)	(0.022)	(0.020)
Rainfall		0.045	0.047	−0.198***	−0.212**
		(0.030)	(0.032)	(0.075)	(0.096)
Rainfall squared		−0.003**	−0.003**	0.016**	0.014
		(0.001)	(0.001)	(0.006)	(0.009)
Observations	15076	15057	11428	11257	11159
Municipality FE	Yes	Yes	Yes	Yes	Yes
Year FE	Yes	Yes	Yes	Yes	Yes
Municipality x year FE	Yes	Yes	Yes	Yes	Yes
Additional controls	No	Yes	Yes	Yes	Yes

Note *** p < 0:01, ** p < 0:05, * p < 0:1. Standard errors in parenthesis are clustered at municipality level. The dependent variable is the number of farm members dedicated to Agriculture, reported in logarithm. Additional controls include household head's age, irrigation, and cultivated land. Data stem from two survey rounds done in 2008 and 2015

Table 9 Weather shocks and labor supply within the farm, by regions

Sub-sample	Dependent variable: number of farm members in agriculture (logged)				
	All (1)	All (2)	Municipalities in		
			Highlands (3)	Valleys (4)	Lowlands (5)
Degree days	−0.002	0.024	0.003	0.015	−0.041
	(0.023)	(0.022)	(0.020)	(0.035)	(0.042)
Harmful degree days	0.013	0.024	0.030	−0.006	−0.061
	(0.021)	(0.021)	(0.025)	(0.032)	(0.045)
Rainfall		0.045	−0.025	−0.130**	−0.122
		(0.030)	(0.126)	(0.052)	(0.088)
Rainfall squared		−0.003**	−0.004	0.004*	0.011**
		(0.001)	(0.012)	(0.002)	(0.005)
Observations	15076	15057	5818	6067	3172
Municipality FE	Yes	Yes	Yes	Yes	Yes
Year FE	Yes	Yes	Yes	Yes	Yes
Municipality x year FE	Yes	Yes	Yes	Yes	Yes
Additional controls	No	Yes	Yes	Yes	Yes

Note *** p < 0:01, ** p < 0:05, * p < 0:1. Standard errors in parenthesis are clustered at municipality level. The dependent variable is the number of farm members dedicated to Agriculture, reported in logarithm. Additional controls include household head's age, irrigation, and cultivated land. Data stem from two survey rounds done in 2008 and 2015

Table 10 Weather shocks and hiring workers, by climate risk areas

Sub-sample	Dependent variable: number of workers hired by the APU (logged)				
	All (1)	All (2)	Municipalities with medium-high risk of		
			Flood (3)	Drought (4)	Frost (5)
Degree days	0.014	0.013	0.006	0.005	0.006
	(0.013)	(0.014)	(0.017)	(0.016)	(0.016)
Harmful degree days	−0.015	−0.022	−0.036**	−0.021	−0.018
	(0.012)	(0.014)	(0.018)	(0.017)	(0.017)
Rainfall		−0.008	−0.006	0.043	0.048
		(0.020)	(0.022)	(0.050)	(0.065)
Rainfall squared		0.001	0.001	−0.004	−0.005
		(0.001)	(0.001)	(0.004)	(0.005)
Observations	15129	15057	11428	11257	11159
Municipality FE	Yes	Yes	Yes	Yes	Yes
Year FE	Yes	Yes	Yes	Yes	Yes
Municipality x year FE	Yes	Yes	Yes	Yes	Yes
Additional controls	No	Yes	Yes	Yes	Yes

Note *** p < 0:01, ** p < 0:05, * p < 0:1. Standard errors in parenthesis are clustered at municipality level. The dependent variable is the number of workers hired by the APU, reported in logarithm. Additional controls include household head's age, irrigation, and cultivated land. Data stem from two survey rounds done in 2008 and 2015

Table 11 Weather shocks and hiring workers, by regions

Sub-sample	Dependent variable: number of workers hired by the APU (logged)				
	All (1)	All (2)	Municipalities in		
			Highlands (3)	Valleys (4)	Lowlands (5)
Degree days	0.014	0.013	−0.016	0.060**	0.089**
	(0.013)	(0.014)	(0.020)	(0.023)	(0.040)
Harmful degree days	−0.015	−0.022	−0.004	−0.034*	−0.017
	(0.012)	(0.014)	(0.041)	(0.019)	(0.050)
Rainfall		−0.008	0.036	−0.048	0.070
		(0.020)	(0.149)	(0.034)	(0.066)
Rainfall squared		0.001	−0.003	0.003**	−0.005
		(0.001)	(0.016)	(0.001)	(0.003)
Observations	15129	15057	5818	6067	3172
Municipality FE	Yes	Yes	Yes	Yes	Yes
Year FE	Yes	Yes	Yes	Yes	Yes
Municipality x year FE	Yes	Yes	Yes	Yes	Yes
Additional controls	No	Yes	Yes	Yes	Yes

Note *** p < 0:01, ** p < 0:05, * p < 0:1. Standard errors in parenthesis are clustered at municipality level. The dependent variable is the number of workers hired by the APU, reported in logarithm. Additional controls include household head's age, irrigation, and cultivated land. Data stem from two survey rounds done in 2008 and 2015

6 Conclusions

In this study we try to understand how agricultural outcomes are affected by weather shocks at the farm-level in Bolivia. In the climate-economy literature this country has not received much of attention, although its geography, its socio-economic conditions and the importance of agriculture to people's wellbeing make it worth study it.

Using micro-data from the Agricultural Survey 2008 and 2015, satellite climate data, and a municipality level fixed effects model, we show that weather shocks, such as extreme temperature and lack of or excess of rainfall have detrimental effects on Bolivian farmers' agricultural yield. Specifically, temperatures that exceed department-specific thresholds decrease yield in between 15 to 19%, depending on the climate risk area the farm is located. The magnitude of the impact of the excess or the scarcity of rainfall on yield is small but negative. Farmers that live in highlands municipalities seem to be more affected by extreme temperature since their yield decreases in about 36%.

In terms of coping actions, we find that cultivating more land for example is not statistically significant. What we do observe is that extreme temperature affects the labor supply decisions inside the farm. The number of members of the farm in agricultural activities increase in 5% with extreme temperature, but only in flood-prone municipalities, in the rest, the estimate is not significant but with the expected sign.

Similarly, farms in flood-prone municipalities decrease in around 4% the number of hired workers when they suffer from extreme temperature.

These coping actions against weather shocks that are observed only for farmers that live in flood-prone municipalities, which compromises most the lowlands region, plus the fact that farmers that live in highlands municipalities are affected the most in terms of yield reductions, lead us to conclude that weather shocks do matter for all farmers and the differences on the magnitude of the impact and the coping actions by farmers in Bolivia are needed to be taken into consideration when implementing policies dedicated to mitigate weather extreme variations.

Appendix

Tables 12 and 13.

Table 12 Weather shocks and yield, by climate risk areas

Sub-sample	Dependent variable: Yield (Logged)				
	All (1)	All (2)	Municipalities with medium-high risk of		
			Flood (3)	Drought (4)	Frost (5)
Precipitations above 1 St. Dev.	0.050	−0.168	−0.135	−0.469**	−0.564**
	(0.038)	(0.136)	(0.139)	(0.192)	(0.266)
Precipitations below 1 St. Dev.	−0.222***	−0.996**	−0.844*	−1.967***	−2.136***
	(0.078)	(0.440)	(0.480)	(0.563)	(0.801)
Rainfall		0.558**	0.474*	1.275***	1.334**
		(0.236)	(0.249)	(0.349)	(0.516)
Rainfall squared		−0.003	−0.003	−0.023	−0.017
		(0.004)	(0.004)	(0.014)	(0.021)
Average day-time temperature (°C)		−0.026	−0.018	0.006	−0.023
		(0.125)	(0.147)	(0.149)	(0.154)
Observations	14635	14562	11011	10941	10908
Municipality FE	Yes	Yes	Yes	Yes	Yes
Year FE	Yes	Yes	Yes	Yes	Yes
Municipality x Year FE	Yes	Yes	Yes	Yes	Yes
Additional controls	No	Yes	Yes	Yes	Yes

Note *** p < 0:01, ** p < 0:05, * p < 0:1. Standard errors in parenthesis are clustered at municipality level. The dependent variable is the average yield, reported in logarithm. Additional controls include household head's age, irrigation, and cultivated land. Data stem from two survey rounds done in 2008 and 2015

Table 13 Weather shocks and value of livestock, by climate risk areas

Sub-sample	Dependent variable: Value of Livestock (Logged)				
	All (1)	All (2)	Municipalities with medium-high risk of		
			Flood (3)	Drought (4)	Frost (5)
Degree days	0.314***	0.112	0.145	0.171*	0.121
	(0.093)	(0.084)	(0.100)	(0.093)	(0.085)
Harmful degree days	−0.197**	−0.090	−0.089	−0.097	0.018
	(0.081)	(0.088)	(0.108)	(0.099)	(0.084)
Rainfall		−0.507***	−0.477***	−0.644	−0.552
		(0.162)	(0.175)	(0.394)	(0.382)
Rainfall squared		0.011	0.010	0.021	0.008
		(0.007)	(0.007)	(0.034)	(0.034)
Observations	12257	12190	9322	9694	9673
Municipality FE	Yes	Yes	Yes	Yes	Yes
Year FE	Yes	Yes	Yes	Yes	Yes
Municipality x year FE	Yes	Yes	Yes	Yes	Yes
Additional controls	No	Yes	Yes	Yes	Yes

Note *** p < 0:01, ** p < 0:05, * p < 0:1. Standard errors in parenthesis are clustered at municipality level. The dependent variable is the value of livestock owned by the APU, reported in logarithm. Additional controls include household head's age, irrigation, and cultivated land. Data stem from two survey rounds done in 2008 and 2015

References

Acevedo S, Mrkaic M, Novta N, Poplawski-Ribeiro M, Pugacheva E, Topalova P (2017) World economic outlook, October 2017, seeking sustainable growth: short-term recovery, long-term challenges, chapter the effects of weather shocks on economic activity: how can low-income countries cope? pp 117–183. International Monetary Fund

Andersen LE and Verner D (2014, November). Social impact of climate change in Bolivia: a municipal level analysis of the effects of recent climate change on life expectancy, consumption, poverty and inequality. Latin Am J Econ Dev 22:49–83

Aragon FM, Oteiza F, Rud JP (2019) Climate change and agriculture: subsistence farmers' response to extreme heat

Castro M (2018) Clusters de calidad de vida y cambio climático en Bolivia: Un análisis espacial multitemporal aplicando sistemas de información geográfica. Latin Am J Econ Dev 29:103–147

Delgadillo MF, Lazo A (2015) Evaluación de daños y pérdidas por eventos climáticos: Bolivia 2013–2014. Technical report, Unidad de Análisis de Políticas Sociales y Económicas (UDAPE)

Dell M, Jones BF, Olken BA (2014). What do we learn from the weather? the new climate-economy literature. J Econ Lit 52(3):740–798

Deschenes O, Greenstone M (2007) The economic impacts of climate change: evidence from agricultural output and random fluctuations in weather. Am Econ Rev 97(1):354–385

Eckstein D, Hutfils M-L, Winges M (2019) Global climate risk index 2019: Who suffers most from extreme weather events? weather-related loss events in 2017 and 1998 to 2017. Technical report, Germanwatch Nord-Sud¨ Initiative eV

Funk C, Peterson P, Landsfeld M, Pedreros D, Verdin J, Shukla S, Husak G, Rowland J, Harrison L, Hoell A et al (2015) The climate hazards infrared precipitation with stations–a new environmental record for monitoring extremes. Scientific data 2(150066)
Gorriti D, Gutierrez S (2014) Metodología para el cálculo del Índice de riesgo municipal con datos del censo 2012. Technical report, World Bank
IPCC (2014) Climate change 2014: synthesis report. contribution of working groups i, ii and iii to the fifth assessment report of the intergovernmental panel on climate change. techreport, IPCC
Schlenker W, Hanemann WM, Fisher AC (2006). The impact of global warming on U.S. agriculture: an econometric analysis of optimal growing conditions. Rev Econ Stat 88(1):113–125
Schlenker W, Roberts MJ (2009). Nonlinear temperature effects indicate severe damages to U.S. crop yields under climate change. Proc Natl Acad Sci 106(37):15594–15598

Pasture Loss Indexed Insurance in Chile

María José Pro, Ibar Silva, Julia Sanz, Juan Carlos Cuevas, and Isaac Maldonado

1 Introduction

1.1 The Chilean Context

Livestock production in Chile has great socioeconomic relevance, with a growing trend that reached 24% of GDP agroforesty and livestock in 2009. The most significant species quantitatively, is the sheep, with a census of over 3.8 million head, raised

We would like to express our gratitude to the Inter-American Development Bank for financing a great part of the actions undertaken in the implementation of this project. Our thanks to the former President of ENESA, Jaime Haddad Sánchez de Cueto, José María García de Francisco, Executive Manager of ENESA and COMSA Executive Manager Ricardo Prado for their institutional support which has made it possible to carry out this initiative. Finally, we would like to thank Fernando Burgaz Moreno for strengthening relations between ENESA and COMSA, during the years he served as Director of ENESA, a period during which the idea of initiating this study was conceived. We also thank the collaboration of José Luis Casanova and Pablo Salvador (LATUV), Marcel Castillo Fuentes and Carolina (INIA-Chile); Elsa Sanchez and Manuel Cardo (AGROSEGURO); Camilo Restrepo and Paula Valdés (COMSA) and Ascension Garcia (ENESA).

M. J. Pro (✉)
ENESA, Ministerio de Agricultura Alimentación Y Medio Ambiente de España, Madrid, Spain
e-mail: mprogonz@magrama.es

I. Silva
Comité Del Seguro Agrícola de Chile, Santiago de Chile, Chile

J. Sanz
University of Valladolid, Valladolid, Spain
e-mail: julia@latuv.uva.es

J. C. Cuevas
Agrupación de Entidades Aseguradoras Del Seguro Agrario En España, Madrid, Spain
e-mail: jccuevas@agroseguro.es

I. Maldonado
Instituto de Investigaciones Agropecuarias, Santiago de Chile, Chile

J. J. Durante and R. Rosillo (eds.), *Natural Disasters and Climate Change*, SpringerBriefs in Economics, https://doi.org/10.1007/978-3-030-43708-4_3

largely in extensive, thanks to a wide area of pasture, which represents 17% of the country land (Ahumada 2004).

Livestock production based on utilization of pastures is exposed to agro climatic factors that make it a very vulnerable sector, especially to drought and other adverse weather conditions, the impact of which reduces the availability of food for animals. For example, in the 2007–2008 season, the Chilean Ministry of Agriculture, through the National Emergency System Agricultural Risk Management and Agro-climate, allocated 16,740 million pesos to mitigate the effects of drought, to which added a loss of revenue from value-added tax (VAT) by decreasing the production estimated at 24,420 million pesos, only in the regions of Atacama and Aysen.

Climate change introduces new uncertainties in livestock production against the need to develop innovative strategies for adaptation, mitigation and risk transfer, among which mechanisms for loss of pasture insurance might be envisaged.

The development of technologies associated with satellite images has opened the way for numerous applications in various fields. Among others, the construction of vegetation indexes as drought indicators has boosted the development of index insurance programs worldwide. In many cases these are just pilot experiences, with limited application (Gommes y Kayitakire 2013). Despite all this, it is worth mentioning the outstanding Spanish program that provides an indexed insurance for the loss of grazing land, which has been on since 2001.

In the case of Chile, analysing the feasibility of an indexed insurance for grazing losses due to adverse weather conditions becomes particularly relevant, since the committee responsible for managing grant programs intended to co-finance agricultural insurance has been lately in favour of promoting the development of parametric insurance to complement traditional yield and damage insurance. Likewise, the Chilean government promotes this instrument because it allows for insurance expansion, reaching a so far unattended segment of 120,000 farms whose main production limitation lies on shortage of winter fodder resulting from drought.

1.2 Purpose and Justification of This Study

Given the above background, a study has been carried out to establish a system of protection against loss of grazing land due to adverse weather, under an experimental type of indexed insurance for the sheep farming regions of Maule and Biobio (Chile). For the assessment of the damage impact the Normalized Difference Vegetation Index (NDVI) measured by means of remote observation will be used, an index that correlates multitude of variables such as vegetation vigor and photosynthetic activity.

The proposed model is based on public and private partnerships, so that private insurance companies offer coverage with the government's support and protection. Thus, economic solvency and legal security of policyholders and insurers are improved and help to strengthen farmers' resilience. They also facilitate access to other financial instruments. All these issues provide opportunities for promotion, innovation and improvement of productivity in the livestock sector with the consequent impact on rural economy as a whole.

Ultimately, loss pastures indexed insurance aims to compensate extensive livestock producers for the increase in food costs in situations of reduced availability of grazing due to bad weather. This insurance may be directed to extensive livestock producers of any species (bovine, caprine, ovine, equine, alpaca, llama, vicuna …).

In order to achieve this aim, analysing the possibilities of insurance is prerequisite. The economic and technical feasibility conditions for implementing the insurance are to be established as well as the overall framework for action and coordination among all stakeholders involved, both public and private. In this sense technology and knowledge transfer plays a fundamental role in monitoring the evolution of vegetation indexes by means of satellite images and their application to indexed insurances.

The model here put forward has been designed taking into account the experience and the lessons learned when implementing the indexed insurance in Spain, adapting it to the particular agricultural and climatic features of the scope (inner dry land areas in the regions of Maule and Biobío).

The design process begins with an agro-climatic characterization of the productive system and the definition of homogeneous areas of pasture in use. Then the historical evolution of NDVI in each of the defined areas is analysed in order to characterize spatial-temporal risk. This information is used to define the insurability conditions of pasture loss due to climatic adversity, specifying levels of coverage, warranty periods, minimum damage threshold, franchises and all the technical aspects needed to design a sustainable, feasible indexed insurance economically acceptable by all the participants involved. Depending on the insurability conditions set, the pricing of risk and an economic evaluation of the implementation of the model is carried out.

Finally it should be noted that the proposed insurance model should be validated and improved on the basis of the results obtained, in order to make it more and more accurate in the application areas and also with a view to expanding its scope as the corresponding actuarial and technical studies are developed, thus progressively generating experiences that facilitate decision making processes on widespread application of the model and encourage the spread of that knowledge to other Latin America countries.

2 Indexed Insurance

2.1 Basic Concepts

Indexed insurance is an innovative risk transfer tool, in which the payment of compensation is triggered when an index, which serves as an indicator of the impact of the risk insured, reaches a certain threshold correlated with the loss suffered by the insured good.

This instrument is used to provide protection against risks that are difficult to objectify, assess and quantify, such as drought and other adverse weather conditions. Drought is an evolutionary risk, with highly variable behavior in different agro-climatic zones, whose direct field assessment would imply high costs and great expertise on the part of qualified technicians.

Therefore, the use of an indirect process that allows for objective, widespread standardized assessment in a large territory is hugely advantageous when designing an insurance.

The choice of procedure to be used in the assessment is essential to ensure the viability and the proper functioning of the insurance, as well as to gain the trust of producers, insurers, reinsurers and governments.

Various types of indexes are used to monitor the evolution of crops and applied to the design of insurances. Those based on climatological data, mainly related to rainfall, are outstanding among them. However, these indices can only be applied in areas where an adequate network of weather stations is available that provide sufficient information, which is rare especially in places with large geographic and climatic diversity.

As opposed to this, the use of indexes calculated on the basis of satellite collected data enables covering large areas and having historical series that allow for risk studies, which are required for the design and the implementation of insurances (Báez 2010).

2.2 *Indexed Insurances Strengths and Weaknesses*

The use of an index in the design of an insurance presents some advantages (Skees 2008) among which the following are outstanding:

- Reduction of moral hazard. Since no direct assessment in the field is done, the practices of farm management applied by the insured do not cause risk worsening.
- Reduction of adverse selection. The guaranteed level for all producers in a region is the same. This prevents adverse selection at the time of the insurance agreement.
- No expenses of expert opinion in the field. Although in the initial stages of implementation of this insurance, setting up checkpoints in the field is recommended to improve the index estimation.
- Establishment of a standardized procedure intended to improve information transparency.

Alongside these strengths, some limitations should also be considered such as:

- Choosing the appropriate index. It is essential to choose an index that is objective and presents a high correlation with previous damage occurring in the field. It should be noted that index insurance is not panacea for widespread application to all kinds of risks and productions, or for all geographic areas.

- Basis risk. It is advisable to consider that there will always be policyholders who will receive lower compensation than the damage suffered and others who will receive higher compensation than the actual loss. This basis risk, however, may be reduced by improving the index estimation.
- Producers' difficulties to understand. Farmers generally have a rather negative perception of agricultural insurance. Alongside this, the technical complexity of indexed insurance and the lack of field damage assessment involve difficulties related to producers' acceptance.
- In the initial implementation stages of indexed insurance significant financial investment on research and development of the model is needed.

2.3 *Vegetation Index and Indexed Insurances*

Many projects based on space observation are aimed to characterize the type, quantity and condition of surface vegetation. This work makes use of different vegetation indexes which are combinations of two or more bands measured with sensors previously attached to different parts of the electromagnetic spectrum where the vegetation has defined characteristics. Thus, vegetation indexes exploit the difference in reflectance presented by the vegetation along the electromagnetic spectrum, for example between the band of red and close infrared, to estimate various characteristics of vegetation cover.

This method is based on spectral reflectance of healthy vegetation and on that of wet and dry soil. Vegetation absorbs almost all incident radiation in the red, with a reflectance below 0.1, i.e. <10% of incident red radiation, while the soils, both wet and dry, reflect almost 30% of the red radiation they obtain from the sun.

By contrast, in the near infrared area, which our eyes do not detect, vegetation has a very high reflectance over 50%, while soils reflect much less, between 20 and 30%, only slightly more than they reflect the red.

Combining the reflectance of the red and close infrared bands, i.e., the sensor measurements in these two bands, it is easy to distinguish the soil covered by vegetation from bare soil: vegetation reflects much close infrared and little red, while the soil, both dry and wet, reflects slightly more infrared than red, though with close to 20–30% moderate values.

Should there be any vegetation problems, its red reflectance increases, that is, it presents more reddish shades. Conversely, close infrared reflectance decreases. This makes the difference between the satellite measurements decrease in the red and infrared ranges. This is due to the fact that, on the one hand infrared decreases and on the other hand red increases. This is the foundation of vegetation indexes, the difference in reflectance between different bands, which accounts for the state of vegetation.

Although these indexes have been used for various studies, the most widely used index, and the one with better results, is the one called NDVI, Normalized Difference Vegetation Index. This index is used to estimate vegetation quantity, quality and development, and it is unanimously acknowledged as the most effective idex for all kinds of situations. NDVI is defined as:

$$NDVI = \frac{(\text{CloseIR} - \text{Red})}{(\text{CloseIR} + \text{Red})}$$

where Red and CloseIR are defined by the spectral reflection measures acquired in the regions of red and near infrared, respectively. These spectral reflections are, in themselves, ratios of the radiation reflected on the incoming radiation in each particular spectral band; therefore they take values ranging from 0 to 1. Thus the NDVI varies between -1 and $+1$: negative values correspond to water or to flooded, snow-covered areas, or to highly watered surfaces, while positive values corresponds to vegetation or to bare soil.

Several investigations have proven that the NDVI index is directly related to the photosynthetic activity of the plant and therefore to the energy absorbed by vegetation cover.

One of the applications in which NDVI results have been most spectacular is in the drought detection. However, it is convenient to admit that this index reflects changes in the vegetative state that may result from other causes than drought. More specifically, NDVI has been applied in Spain since 2001 as a benchmark for the "Insurance of grazing loss due to climatic adversity" with very good results.

2.4 NDVI Data Gathering

NDVI data are drawn from the MODIS (Spectro-radiometer for images Moderate Resolution) sensor. This sensor flies aboard two platforms, Terra and Aqua, which belong to the NASA EOS constellation (Earth Observation Systems) whose main goal is the continuous observation of global changes, including an integrated study of the atmosphere, the oceans and the land surface. Data have high radiometric sensitivity, good geometrical quality (especially the MODIS sensor aboard Terra satellite), high timing resolution, and they are distributed free of charge through various Internet webpages.

The MODIS Land Science Team (Modland) is responsible for developing methods and the algorithms that generate products on land cover, such as NDVI used in this study.

To be more specific, we have drawn all historical data from the NASA archives in connection with NASA's EOSDIS distribution platform (NASA's Earth Observation System Data and Information System) Reverb ECHO. This MOD13Q1 product contains NDVI in HDF format, apart from other information, such as vegetation index, the reflectance of some bands and a quality indicator for each pixel and the viewing angles and sun inclination. It is generated every 16 days and its spatial resolution is 250 m. For the purposes of this study, only the values that have passed the quality filter and with which the evolution of NDVI is obtained are retained, thus preventing pollution generated by clouds or other issues that may mask the results by adding noise to the process.

2.5 Marking Out Homogeneous Pasture Areas

Designing an indexed pasture drought insurance requires to establish an area where the vegetation reflectance integrated behaviour is able to detect behavioral changes that allow for the association of a level of damage that may be considered critical and which has adversely affected the productive expression of the herbaceous cover analyzed. For insurance purposes these areas are referred to as "homogeneous grazing areas".

The result of surveys of producers and experts led to observing that there is great diversity management practices in the use of pasture. Therefore, in order to define homogeneous areas where to implement both the insurance and the analysis of the satellite image information, each of the municipalities in the area of dry land inside the regions of Maule and Biobío were considered as homogeneous grazing areas. In the present study, 8 municipalities in the Region of Maule and 14 municipalities in the Region of Biobío circumscribe so-called inner rainfed land zone.

With a view to processing MODIS satellite information that allows to track the evolution of NDVI in each of the communities along the time series 2000–2012 auxiliary information was used like the digital map of land use, information on political and administrative divisions, including district census tracts, environmental homogeneous maps, and the 1999 survey of the National Forestry Corporation (CONAF). An adjustment has been made trying to identify the areas where there has been a change in land use, through the analysis of main components (MCA), which has been run by the INIA- team Chile.

With this, a first version was generated of maps of homogeneous grazing areas, extracting and contrasting soil areas covered by grassland (annual pasture perennial grassland, prairie, scrub, scrub-grassland, open scrubland, very open scrub, dense scrub pasture, semi-dense scrub, semi-dense tree scrub, and open tree scrubland as shown on Map 1.

3 NDVI Reference Building Process

After selecting the grazing area, we proceeded to analyze the NDVI evolution in all communes in the dry land inner areas of the regions of Maule and Biobío, calculated

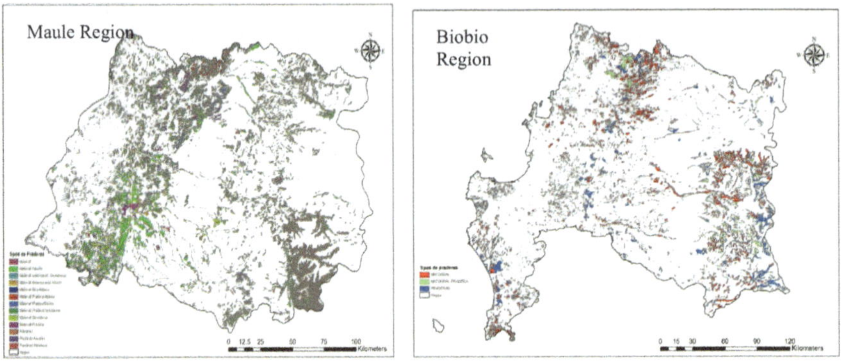

Map 1 Initial map of homogeneous pasture areas

on the basis of the historical 2000–2012 series MODIS images. Additionally, climatological information has been used to validate the behavior of the NDVI evolution curves and to define the period of greatest pasture sensitivity to climate events.

The essential element to obtain an accurate NDVI Reference Curve, in a homogeneous grazing area, is the proper selection of grass types to be considered. Naturally, in case of drought, all types of grass will suffer, as they are in the same area. But, as the detection of drought, from the viewpoint of insurance is obtained by comparing the current NDVI with the reference NDVI, this comparison will necessarily correspond to very similar changes, i.e. the NDVI to be compared and the Reference NDVI correspond to a grass pasture or set of grass pastures which evolve harmoniously. This comparison should be free from bias, and also from the influence of pasture having different peak or minimum times or behaving differently.

3.1 Average NDVI Estimate

Once NDVI filtered compounds and the corresponding crop use masks are available, we obtain the average NDVI for the pixels included within each mask generated from the vectors and for each of the images from 2000 to 2012.

The result of this process is an NDVI average value for each mask of crop use and for each 16 days compound. The NDVI evolution curve consists of 23 values per year for the period of 13 years. This curve gives an idea of NDVI behavior and of its variability for each use thanks to the associated standard deviation value.

To reduce the influence of likely bias caused by contamination in the image, a smoothing on each of these changes is performed. This pollution results from many causes: sun lighting differences or variation in satellite viewing angles, due to variability in the time at which the satellite overpasses and to the logic annual evolution of the position of the sun; also the possibility of very faint mist that do not affect too much the NDVI to eliminate it in terms of quality control; the slope of the terrain and many others. Figure 1 shows the example of the commune of Ninhue after application of smoothing.

Once obtained the curves showing the temporal evolution of NDVI for each type of pasture in each of the municipalities considered, the next step is to find an "Average Curve" or "Reference Curve" for the behavior of NDVI in each of the areas. That is, a curve to be used as the reference to see whether the behavior of the grass in the current year corresponds to what was expected or not. This "Reference Curve" is a key point in the implementation of pasture insurance.

For a curve showing the NDVI annual cycle, smoothed NDVI values are averaged for each commune and each soil use corresponding to each period of 16 days in all the historical series dealt with. In addition to the average made, the standard deviations for each point are also calculated. Thus an average curve—the "Reference Curve" of NDVI evolution in each of the areas and for each type of grass, along with the standard deviation of each of the 23 points that shape the curve—is obtained. Figure 2 shows, for example, the NDVI evolution curve and the average year standard deviation corresponding to the commune of Ninhue in Biobío.

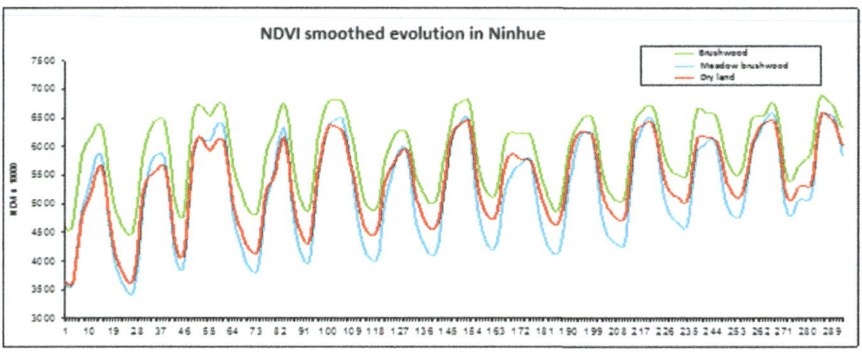

Fig. 1 NDVI smoothed evolution in Ninhue, Biobío. *Source* LATUV

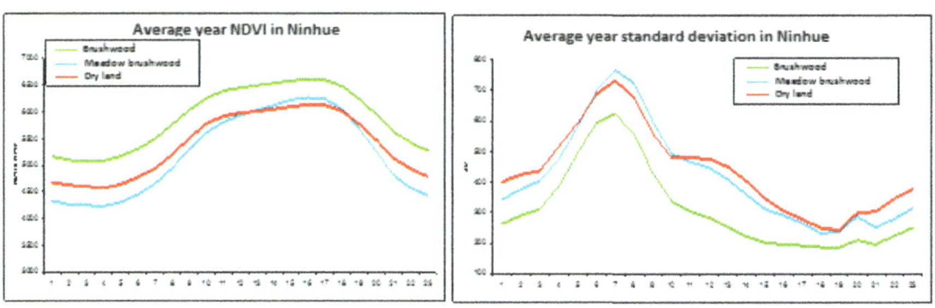

Fig. 2 NDVI evolution and average year standard deviation. Ninhue-Biobío. *Source* LATUV

The above curves are considered as the "Reference curves for each type of grass as they synthesise the average evolution of the grass along the 13 year period considered.

3.2 Vegetable Cover Selection Criteria

Once calculated the reference curves for each of the surfaces of pasture use of in each of the communes and for the entire historical series, it is still necessary to decide what types of grass should be considered representative of a particular commune. Among the criteria for selecting the pasture cover to be used in the construction of the Reference Curve the following are outstanding:

- The most abundant grass should be included in the Reference Curve.
- Should there be other types of grass that match their evolution with that of the most abundant type of grass, they may be added in the calculation of the Reference Curve. It will be necessary to pay attention to the shape of the evolution curve, to the time of their maximum and minimum values, which must be coincident with that of the reference, to its growth rate, defined by the slope of the NDVI curve, etc.
- It is also important to consider the spatial distribution of grazing pasture; those included in the Reference Curve should cover the entire commune.
- Some grasses are more sensitive to the moment of NDVI of minimum value and, if possible, they should be incorporated into the Reference Curve. These and other issues that may arise in specific situations, have to be supplemented with information by experts who are familiar with the soil.

Once the pasture use surfaces to be included in the Reference Curve for each of the municipalities have been agreed on, a weighted average for all the grass in each commune, including the most representative ones, is made. The weight of each type of grass is established according to the number of pixels that are included in each, i.e. it is a linear weighting. This is the reference curve to be used for insurance purposes in each of the municipalities. For example, Fig. 3 shows the NDVI smoothed average year evolution for Ninhue in Biobío.

3.3 NDVI Validation for Insurance Purposes

Before dealing with the risk analysis, it is essential to have additional information to verify the suitability of the construction process of NDVI curves and whether they adequately reflect the behavior of the different types of "pasture" that are normally used by livestock in different homogeneous grazing areas.

To test the consistency of the NDVI evolution of the studied areas, we used a rainfall study conducted by the National Agricultural and Animal Production Research

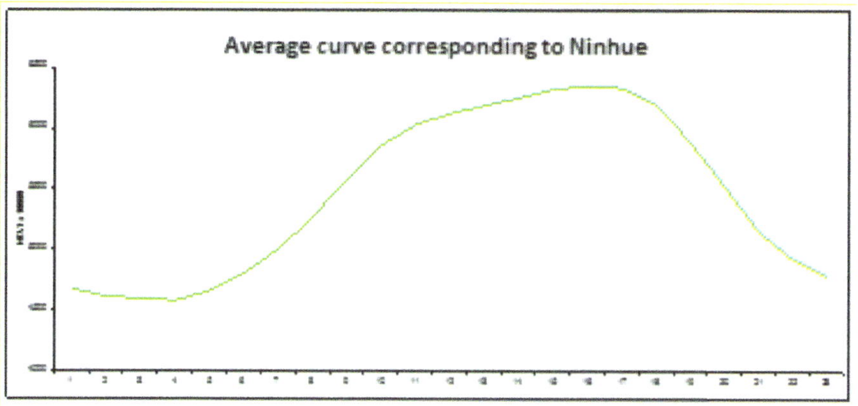

Fig. 3 Smoothed average NDVI evolution in Ninhue, Biobío. *Source* LATUV

Institute of Chile (INIA), on the basis of which two 'good' years and two "bad" years have been fixed for each of the regions under study.

Then AGROSEGURO, SA technical team, have been analysing during the chosen years, the evolution of NDVI in the sensitive period. This "sensitive period" ranging from 1 August and 30 November, was defined on the basis of the analysis of the evolution of the benchmark indexes according to the work of the LATUV—Spain team and to an extensive collection of information (Figueroa 1986; Del Pozo and del Canto 1999; Avedaño et al. 2004) and interviews with experts in agriculture and livestock in the regions of Maule and Biobío, the latter information being obtained mainly through the INIA-Chile team.

After analyzing internal consistency of NDVI evolution curves and their contrast with the rainfall information available, we may conclude that the NDVI evolution curves in the two regions show high consistency and can be used in monitoring the pasture evolution for insurance purposes.

4 Time and Space Risk Identification

Once the NDVI evolution reference curves for each of the communes and the evolution curve of the standard deviation have been built, an initial estimation has been made of the risk of pasture loss due to weather adversity. For the risk characterization the following formula, which will be subsequently used for safe insurance operation purposes, has been used:

where

$$NVDI_g = \left(\bar{X} - (a\delta)\right)$$

Table 1 Initial risk estimation

Model	Region	Risk level (%)
Model 1: Annual guarantee/MMC 5 fortnights	Maule	21,11
	Biobío	19,31
Model 2: 4 months guarantee/MMC 3 fortnights	Maule	17,43
	Biobío	14,92

Source AGROSEGURO, S.A

$NVDI_g$ = is the guaranteed NVDI value

$\bar{X}$ = is the reference vegetations index for each 23 annual periods of 16 days

a = Multiplying coefficient

(δ) = Standard deviation.

Two criteria for risk quantification were applied in this case: the temporal dimension of the phenomenon and the base period for quantification. To do this, two models were designed on the basis of both criteria: "guarantee period" (sensitive period) and the "minimum compensable loss" (minimum number of fortnights below guaranteed level to activate the payment of compensation).

(i) Model 1: The study of one year (23 fortnights) with accumulation of 5 or more fortnights throughout the whole year below the guaranteed level (mínimum compensable claim—MCC). Therefore 21.74% of the fortnights would need to be stricken for entitlement to compensation.

(ii) Model 2: The study of the period running from 1 August to 30 November (8 fortnights), with accumulation of three fortnights below the guaranteed level (mínimum compensable claim—MCC). Therefore 37.5% of the fortnights would have to be stricken for entitlement to compensation.

The study, carried out with a coverage level defined by $a = 1$, revealed high sensitivity to index fluctuation. Therefore, we opted for a threshold coverage defined by the multiplier $a = 1.2$, for an initial estimation of risk probability—after weighting the estimated risk in each commune, in terms of the area affected. This involved identifying an affected area with the chosen vegetal groundcover area in each commune for the construction of NDVI evolution curves of surface. On the basis of these criteria, the resulting risk probability values are shown in Table 1.

5 Definition of Risk Insurability Conditions

After choosing the type of index to be used and after defining the area of action for designing the insurance, it is convenient to specify the basic aspects of the conditions on which the insurance is technically feasible as well as the actuarial aspects that define risk pricing without forgetting fundamental operational aspects for the implementation and management of the insurance, including institutional coordination of public and private sectors, which is essential to ensure the sustainability of the model. The concept of compensable loss and the compensation amounts must set values that involve levels of acceptable risk for all parties.

5.1 Basic Points of Insurance Conditions

The loss of pasture indexed insurance is intended to compensate sheep producers in extensive regime for lack of pasture due to adverse weather conditions. In the context of this project, the implementation of a pilot experiment for sheep farms in extensive communes of the inner dry regions of Maule and Biobío is put forward. This domain may be extended in the future to other livestock species and other areas as experience is gained and as the related actuarial technical studies are carried out.

In this application scope, homogeneous grazing areas are defined to assess and monitor different vegetation indexes. For this purpose, in the particular case of this pilot project, and according to the information available in the technical report prepared under the project, by consensus between the Chilean and the Spanish teams, each commune is considered as a homogeneous area of different grazing.

The periods in which farmers have natural pastures have defined (Díaz-Rossell 2012; Ovalle et al. 1990; Ovalle et al. 1987; Beranguer and Micol 1981) and so have the parts of the year which, according to historical data, have had to face adverse climatic events that have drastically reduced the availability of pasture. The highest vulnerability period due to lack of pasture is concentrated in the months of August to November when the sheep of the areas under study mainly feed on grass eaten on the spot. Hence the proposal to extend coverage during the "sensitive period", which begins on August 1 and ends on November 30 each calendar year. This means 120 days protection.

In the proposed indexed insurance, forage availability is determined by tracking the NDVI for each homogeneous grazing area. So, for each homogeneous area, the current vegetation index (for the current year) is compared to a reference vegetation index (average value calculated with the 2000–2012 series). When each fortnight (NDVI-A) current vegetation index, defined for insurance purposes, is below the guaranteed (NDVI-G) index the availability of pasture is considered to have been reduced. The guaranteed vegetation index may vary depending on the coverage extent that producers are willing to take. Therefore, when developing the risk analysis work and the insurance pricing, different simulations have been made with various guaranteed vegetation index values.

The guarantees have been aimed at providing additional funds to farmers for the purchase of fodder, which they must dramatically purchase to meet the nutritional needs of the sheep, in case of shortage of pasture. It is therefore considered that the insurance provides an offset for the increased expenses needed to supply the necessary fodder for the sheep. Therefore, a value of supplemental food ration per animal per day has been set on the basis of data provided by INIA, by the Chilean Committee of Agricultural Insurance (COMSA) and the stock breeders in the area. Calculating that the value of the food supplement would be $140 Chilean per reproductive animal per day. To simplify the insurance management the compensation includes the proportion corresponding to the food supply for the offspring, thereby establishing the insurance value in $ 180 Chilean per animal per day. This value may be revised at the time of insurance application, with the consequent impact on the final cost of the same.

The insured amount, for insurance purposes, results from multiplying the number of breeding animals declared by the policyholder when making the insurance declaration and the unit price of the insurance, all of which multiplied by the number of coverage days. The insured sum is set at 100% of the insured amount. The insured cattle breeder shall be entitled to financial compensation when the index of current vegetation of the homogeneous zone in which their property is located is below the guaranteed vegetation index, provided the minimum compensable claim is exceeded.

The minimum compensable loss or minimum damage threshold refers to the minimum amount that must reach the damage caused by a risk for which the insured is entitled to receive compensation. According to studies made for the design of such indexed insurance, order to consider a claim as compensable, it is recommended that rate of current vegetation of the corresponding homogeneous area should be smaller than the guaranteed vegetation index for at least three fortnights included in the coverage period.

To be entitled to compensation policyholders must meet a few minimum conditions of animal management. These conditions will be defined by the official responsible for the sheep sector in the Ministry of Agriculture of Chile and they shall be consistent with the most common farming practices in the area.

In case of compensable loss, when calculating the final compensation, a franchise will be deducted. The application of a franchise involves the assumption of risk shared by both insured and insurer. In the corresponding risk rating study, in order to reduce the cost of the insurance, various simulations have been made applying different franchise levels to the risk premiums calculation.

Regarding the subscription period during which the insurance may be arranged, this period of time must avoid any overlapping with the period of risk or with a clear perception of their occurrence. Therefore, it is recommended that the subscription period should end one month before the start of the coverage period. For example, if the insurance starts its guarantees on August 1, June 30 is proposed as the final date of the subscription period.

To facilitate access to information on the state of pastures for insurance purposes, it is recommended that the public body responsible for agricultural insurance should communicate the evolution of the current vegetation index in each of the homogeneous grazing zones, through the institutional website. The information published will result from processing satellite data, according to the protocol established for insurance purposes by the officially appointed institution, probably the National Agricultural and livestock Research Institute of Chile.

5.2 Actuarial Aspects

Given the technical conditions defined in the previous section, guidance is shown on the estimation of the premium for the application of the pasture loss indexed insurance for sheep in the regions of Maule and Biobío in Chile. This analysis, done by the AGROSEGURO, SA, technical team presents several scenarios in order to provide elements that facilitate decision making at the time of marketing the product.

A pricing simulation has been made considering different levels of the guaranteed vegetation index to identify the phenomena that can be classified as extraordinary.

This clearly shows how, as the distance increases between the reference vegetation index and the guaranteed vegetation index the number of compensable communes is gradually reduced, and the number of years during which compensable losses occur.

$a = 1.2$ is the maximum value of the multiplier factor of the standard deviation used to define the guaranteed vegetation index, which reflects widespread losses in all the communes along he two "bad" years.

Figure 4 displays the compensable distribution years and the number of communes entitled to compensation.

The graphs above clearly show the reductive effect on the expected risk level after increasing the standard deviation multiplying factor (Fig. 5).

In any case, the cost of ensuring such protection levels can be reduced by applying a franchise, i.e. a reduction factor of the compensation received by the policyholder thus being assigned to the latter a percentage of insured losses.

The application of a franchise, along with reducing the insurance cost, promotes responsibility in the risk management being shared by both the policyholder and the insurer. This contributes to minimizing moral hazard and encourages the insured to continue implementing the appropriate management practice. To assess the impact of the application of the franchise on the risk premium of the insurance several estimates, have been made with different levels of absolute franchise (0, 10, 20, 30%). Figure 6 presents the results of applying various franchise percentages, which show its impact on the risk premium, for the different levels of insurance presented.

Considering the purchasing power of the sector and the experience of the COMSA team, the reference value recommended for the implementation of the insurance

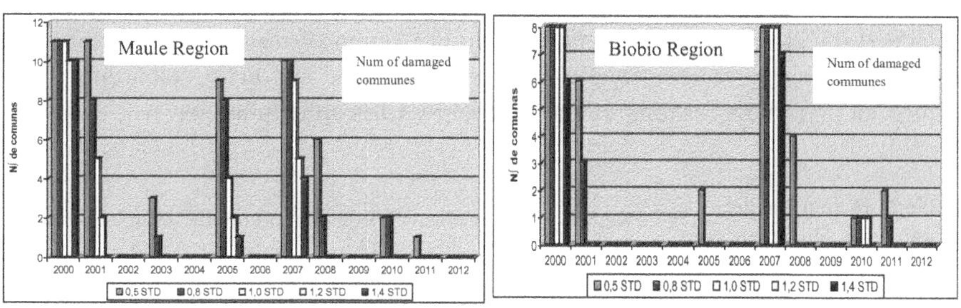

Fig. 4 Damaged comunes versus coverage level. *Source* AGROSEGURO, S.A

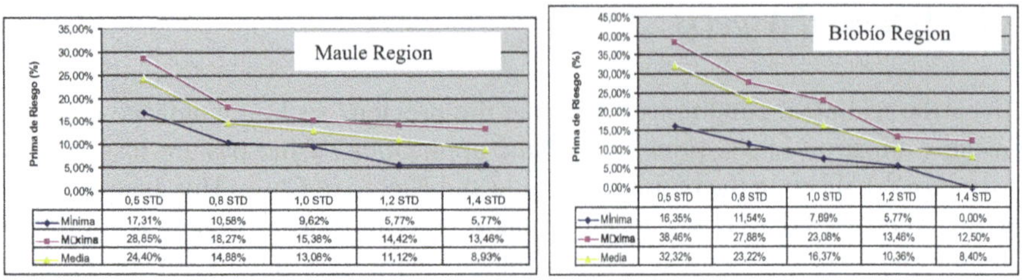

Fig. 5 Risk Premium versus Coverage level. (*Source* AGROSEGURO, S.A). Fuente: AGROSEGURO, S.A

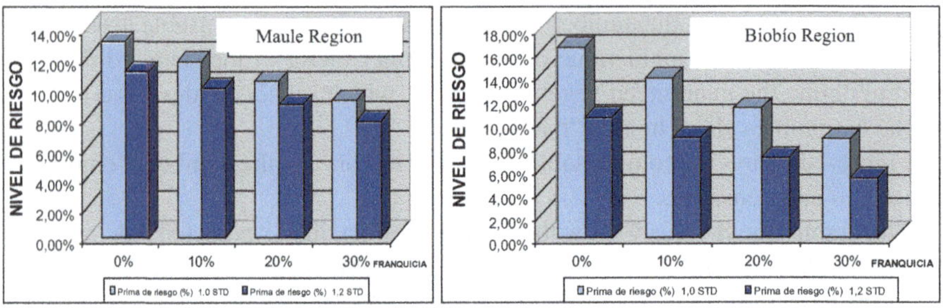

Fig. 6 Risk prime versus guarantee level and franchise. *Source* AGROSEGURO, S.A

would correspond to a guaranteed level defined by $a = 1.2$ and 20% franchise, which for the coverage period ranging from August 1 to November 30 with minimum 3 fortnights compensable claim would result in a risk premium of 8.89% in the region of Maule and of 6.90% in the region of Biobío.

From these values, and taking into account the actual situation for the development of the insurance, the likely advantages and disadvantages of establishing either a single premium, or one with different values per region, as well as the criteria for establishing premiums must be assessed. If the premium is fixed on the basis of the distribution of the number of sheep head in both regions, the final value is shown in Table 2.

We should bear in mind that this refers to the premium base, to which should be added the cost of management and administration of insurance companies and agents' commissions, security fees and/or reinsurance and taxes, according to the comercial premiums formulation criteria set by Chilean regulations.

Table 2 Homogeneous single scope. Risk premium

Region	Sheep census	Regional premium (%)	Single premium (%)
Maule	83.233	8,89	8,47
Biobío	22.534	6,99	

Source AGROSEGURO, S.A

5.3 Economic Assessment

Obviously, the economic evaluation of the insurance implementation will depend on the specific level of coverage chosen, on the value of the food ration defined for the purposes of calculating the insured amount and the compensation payment, on the minimum compensable loss, on the percentage of franchise and, finally, on the chosen area.

In order to estimate the insurance application cost in the upland dry comunes within the regions of Maule and Biobío in Chile, the insurance operating parameters that are economically most viable have been picked out from all the simulated scenarios taking into account the approximate level of risk provided by the COMSA team. Therefore, the economic estimate presented below is based on the following parameters:

Warranty: Compensation for unavailability of grass, assigning a value of the diet of breeding animals corresponding to 180 Chilean pesos per day per animal.
Coverage level: Equivalent to a guaranteed vegetation index resulting from applying a value of a = 1.2 to the multiplier factor of the standard deviation.
Period of Coverage: The greatest vulnerability period estimated, and therefore the guarantee period extends from August 1 to November 30 (120 days).
Minimum compensable loss or minimum damage threshold: Equivalent to three or more cumulative fortnights in which the current vegetation index is located below the guaranteed vegetation index.
Compensation: In case of compensable loss a compensation of 100% of the value of the diet is established per breeding animal per day.
Number of insurable animals: This is the number of breeding animals according to the VII National Census (INE 2011). For insurance purposes, the following types are to be considered as breeding animals: male and female sheep, rams, capons and ewes.

Previous experience in different countries and professional branches shows that when a new insurance is launched, it is difficult to reach insurance for 100% of the animals during the first year. It is logical to expect a gradual process of recruitment which starts slowly. Therefore, Table 3 presents in detail the result of estimating the economic assessment of the implementation of the insurance under the above conditions and considering insurance for 20% of the animals (Table 3).

As mentioned above, the estimates of insurance application cost presented in this section have been made only the premium base. At the time of application of the model, it would be necessary to add the cost of management and administration of the insurance companies and other fees that may be established in compliance with the current Chilean legal regulations. Finally it would be necessary to apply the criteria for granting subsidies to the premium to know the final cost of the insurance corresponding to the farmers.

Table 3 Economic assessment for a 120 day guaranty with 20% implementation

Zone	Insurable animals Num	Price per unit ($*/animal and day)	Guaranty period (Num of days)	Amount of the risk (Millions $*)	Basic premium risk (%)	Insurance cost (Millions $*)
R. Maule	16.647	180	120	360	8,89	31,97
R. Biobío	4.507	180	120	97	6,90	6,72
Single scope	21.153	180	120	457	8,47	38,68

Source ENESA
*$ equivalent to Chilean peso

6 Conclusion

The last decade has witnessed the development of an important scientific body focused on evaluating the potential of indexed insurance and on drawing lessons from the first pilot program experiments implemented in different parts of the world (Leeuwet 2014, Miranda and Farrin 2012; Báez 2010; Barnet et al. 2008; Skees 2008).

In some countries weather data-based indexes have been used to design insurance; in others satellite images have been used. Regarding the use of remote sensing, the application and usefulness of NDVI is clearly confirmed in monitoring crop vegetative state. This index enables charging climate risks in the insurance and is especially useful when the weather information is scarce. Therefore, this index, has been in operation in the application of an indexed insurance to provide coverage against loss of pasture for livestock in Spain since 2001.

In the case of Chile, the government shows a strong political will to develop and extend insurance that allow for strengthening the resilience of the population engaged in agricultural and livestock activities. In particular, in the last few years they are developing a special interest in indexed insurance, evidenced by the positive attitude of the Superintendencia de Valores y Seguros. Also, some private insurance companies operating in Chile, show an interest in selling a pasture loss indexed insurance.

Therefore, the knowledge and technology transfer carried out by Spanish experts in remote sensing for agricultural insurance has contributed to strengthening research institutions as well as other institutions linked to agricultural and livestock insurance in Chile. Likewise, the necessary technical foundations have been laid to experimentally implement an indexed insurance in order to cover pasture loss caused by adverse weather in sheep farming in the regions of Maule and Biobío.

This insurance must be inserted in an overall framework of risk management strategies and its widespread implementation involves a long, slow development process as well as high cost, especially in the initial stages. The supervision, regulation and coordination role of the government should not be ignored as it is essential to ensure the proper functioning of the model and its sustainability in the medium and long term.

The results of the experience obtained with the pilot program will facilitate decision-making on the widespread application of the model to other livestock species and in different geographical areas of the country and they will promote the extension of the knowledge acquired in the field of agricultural and livestock risk management by means of satellite images to other agricultural sectors and to the improvement of environmental management. It may also provide a focal point for the transfer of knowledge to other countries in Latin America and the Caribbean.

References

Ahumada M (2004) Las praderas en Chile y su relación con el sector. Boletín del Departamento de Protección de los Recursos Naturales Renovables. Servicio Agrícola y Ganadero. Santiago, Chile

Avendaño J, Fernández F, Ovalle C, Blu F (2004) Ovinos alimentados con raciones que incluyen tagasaste (*Chamaecytisus proliferus* subsp. palmensis) en reemplazo de heno de alfalfa. II. Digestibilidad y consumo de nutrientes. Agricultura Técnica 64:271–279

Báez K (2010) El Potencial del Seguro Indexado en Chile: Una Aplicación a la Gestión del Riesgo de Sequía en Pastos. Universidad Politécnica de Madrid. Escuela de Ingenieros Agrónomos, Departamento de Economía y Ciencias Sociales Agrarias

Barnett B, Barrett C, Skees J (2008) Poverty traps and index-based risk transfer products. World Dev 36(10):1766–1785

Beranger C, Micol D (1981) Utilisation de l´herbe par les bovins au paturage. Importance du chargement et du mode d´exploitation. Fourrages 85:73–93

Del Pozo A, Del Canto P (1999) Zonas agroclimáticas y sistemas productivos de la VII y VII región. Instituto de Investigaciones Agropecuarias, Centro de Investigaciones Quilamapu. Serie Quilamapu Nº 113. 89

Díaz-Rossell R (2012) Síntesis y Análisis Integrado de la Información de los Sistemas de Agricultura Familiar y sus Características Frente al Cambio Climático. Informe IICA-Procisur del proyecto Estrategias de Extensión: Los agricultores familiares y su adaptación al CC en territorios seleccionados del Cono Sur. Argentina, Chile y Uruguay

Figueroa M (1986) Recursos forrajeros utilizados en producción ovina. III Zona centro-sur. In: García G (ed) producción ovina. Universidad de Chile, Facultad de Ciencias Agrarias y Forestales, Santiago, pp 81–92

Instituto Nacional de Estadística, INE (2011) Encuesta de ganado ovino 2010. Santiago de Chile

Leeuw J, Vrieling A, Shee A, Atzberger C, Hadgu KM, Biradar ChM, Keah H, Turvey C (2014) The potential and uptake of remote sensing in insurance: a review. Remote Sens 6:10888–10912. https://doi.org/10.3390/rs61110888

Miranda MJ, Farrin K (2012) Index insurance for developing countries. Appl Econ Perspect Policy 34:391–427

Ovalle C, Avendaño J, Acuña H, Soto P (1987) La carga animal con ovinos en el espinal de la zona mediterránea subhúmeda. III. Comportamiento animal. Agricultura Técnica 47:211–218

Ovalle C, Aronson J, del Pozo A, Avendaño J (1990) The espinal: agroforestry systems of the mediterranean-type climate region of Chile. Agrofor Syst 10:213–239

Skees J (2008) Innovations in index insurance for the poor in lower income countries. Agric Resour Econ Rev 37(1):1–15

The manufacturer's authorised representative in the EU is Springer
Nature Customer Service Centre GmbH, Europaplatz 3, 69115 Heidelberg,
Germany. If you have any concerns regarding our products, please
contact ProductSafety@springernature.com

Printed and bound by CPI Group (UK) Ltd, Croydon, CR0 4YY
30/09/2026
02225467-0001